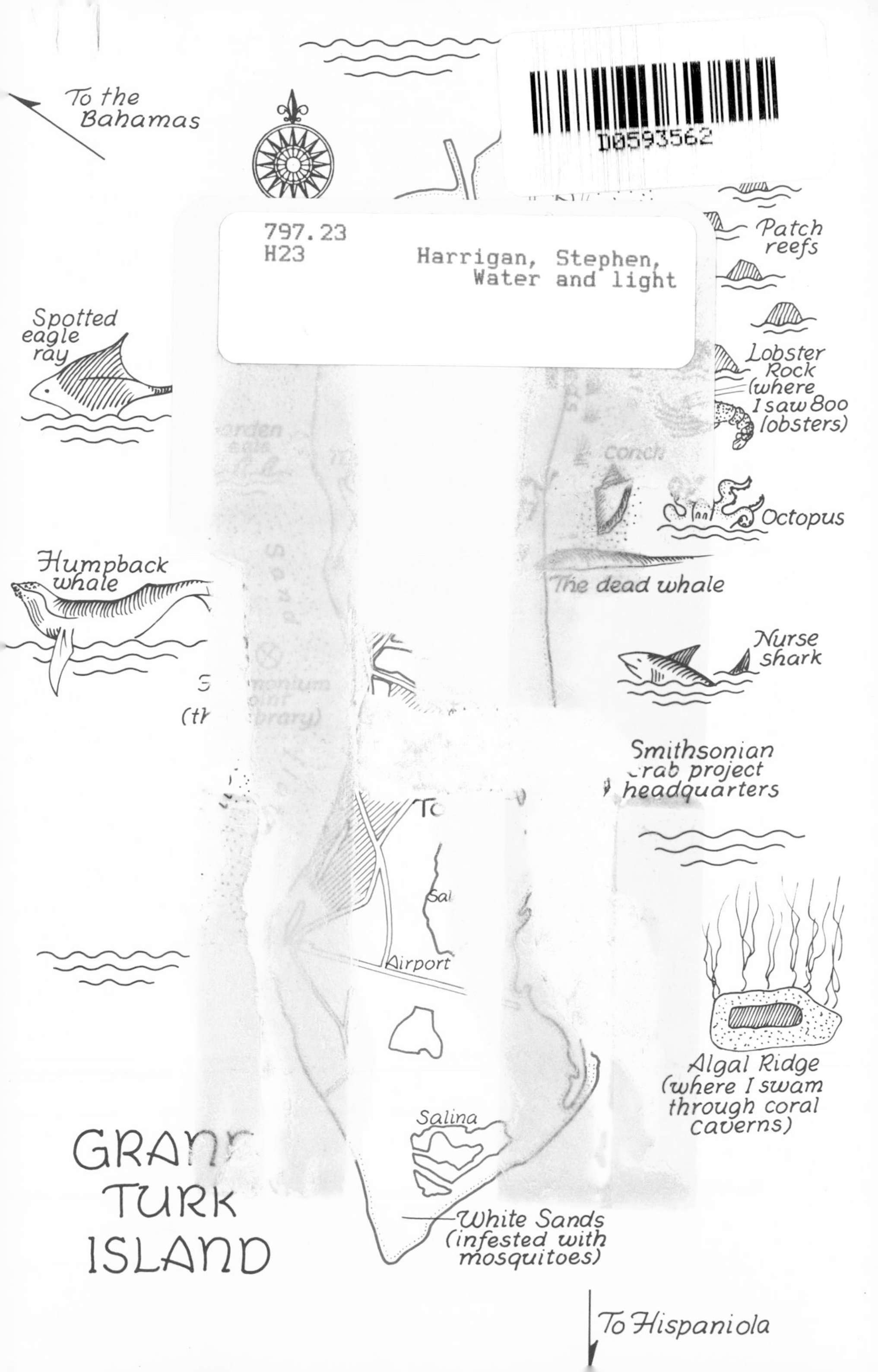

To the Bahamas
Patch reefs
Spotted eagle ray
Lobster Rock (where I saw 800 lobsters)
conch
Octopus
Humpback whale
The dead whale
Nurse shark
Smithsonian crab project headquarters
Airport
Algal Ridge (where I swam through coral caverns)
Salina
GRAND TURK ISLAND
White Sands (infested with mosquitoes)
To Hispaniola

Water and Light

Books by Stephen Harrigan

Aransas

Jacob's Well

A Natural State

Water and Light:
A Diver's Journey to a Coral Reef

Stephen Harrigan

Water and Light

A Diver's Journey to a Coral Reef

HOUGHTON MIFFLIN COMPANY
Boston / New York / London 1992

For information about permission to reproduce selections from this book, write to Permissions, Houghton Mifflin Company, 215 Park Avenue South, New York, New York 10003.

Library of Congress Cataloging-in-Publication Data

Harrigan, Stephen, date.
Water and light / a diver's journey to a coral reef / Stephen Harrigan.
p. cm.
Includes bibliographical references and index.
ISBN 0-395-46558-3
1. Underwater exploration. 2. Diving. 3. Coral reef ecology.
I. Title.
GC65.H34 1992 91-45623
797.2'3—dc20 CIP

Printed in the United States of America

Book design by Robert Overholtzer
Endpaper map by Jacques Chazaud

BP 10 9 8 7 6 5 4 3 2 1

Portions of this book have been published in several periodicals: chapter 1 appeared in *Texas Monthly;* parts of chapters 2 and 10 appeared in *Condé Nast Traveler;* chapter 3 appeared in *Audubon;* parts of chapter 5 appeared in *Icarus;* part of chapter 6 appeared in *Special Report;* and chapter 13 appeared in *Outside.*

For Marjorie, Dorothy, and Charlotte

Contents

It is quite conceivable that underwater man will be spiritually transformed by his activity, that from his intercourse with the sea he will receive an unexpected gift: a certain wisdom, a different way of thinking, judging and making decisions.

Jean-Albert Foëx, *The Underwater Man*

1

My Underwater Self

I no longer live near the ocean. The nearest salt water now is two hundred miles away across a flat coastal plain whose bedrock was formed from the muck and calcium of an ancient sea. But when I was a boy I lived on the coast itself, and I went to sleep every night with my mind peacefully roving through the dark waters of the bay.

The bay was murky, but in my dreams the water became so clear I could feel my eyes straining from the effort to extend their range, to locate some finite point in that endless crystal void. The creatures I saw gliding about underwater were always mysteriously benevolent. They were not fish usually, but half-glimpsed amalgams of real and imagined animals, adapted — as I apparently was — for underwater life. They had been waiting for me to appear. The water's sudden clarity seemed to have roused them, as if until now they had been physically trapped in their gloomy element like prehistoric animals in a peat bog. I felt released too, beyond the reach of wakeful caution, beyond the jurisdiction of physical laws. I could breathe, I could range wherever my will would take me, soaring along the contours of the sea bottom or spiraling up toward the surface, into the high altitudes of the ocean atmosphere.

All my life I have dreamed one variant or another of that dream. I have had a passion to be underwater. How this passion developed I'm not sure, but I remember the longing I felt — the brutal, unappeasable longing of a very young child — when my mother used to read to me, night after night, a story called *The Water Babies*.

The Water Babies is a novel for children written in 1862 by a strange, sex-tormented Victorian cleric named Charles Kingsley. According to his biographer, Susan Chitty, Kingsley "could only accept the idea of carnal relations with his wife once he had convinced himself that the body was holy and the act of sex a sacrament in which he was the priest and his partner the victim." Kingsley sorted through his obsessions by writing verse and best-selling novels and by producing a series of disturbing drawings that depicted him and his wife, Fanny, in rapturous postures of self-mortification — drawings that, according to Nathaniel Hawthorne, "no pure man could have made or allowed himself to look at."

And yet purity was Kingsley's lifelong ideal. *The Water Babies* is the story of a poor and abused boy named Tom, whose work as a chimney sweep has left him habitually covered with grime. While servicing the chimneys of a country gentleman's estate, he finds himself in the presence of a sleeping girl whose angelic cleanliness makes him quake with desire and shame. When she awakes and sees him by her bed she screams. He flees from the house, telling himself, "I must be clean, I must be clean." Finally he comes to a clear brook. Entering the water, he falls into "the quietest, sunniest, cosiest sleep that ever he had in his life" and wakes up reborn as a water baby, a little naked human form four inches long, "with a pretty lace collar of gills." In this form Tom goes through a series of adventures involving a courtly salmon, a ferocious mother otter, and a dimwitted lobster. The story grows increasingly weird as its author's great throbbing themes of purgative redemption and "muscular Christianity" crowd out any hope of narrative coherence.

The version of *The Water Babies* that was read to me as a child was much simplified, a heavily illustrated condensation of the story in rhyming quatrains that appeared in a popular series of children's literature called *My Book House*. I still have that volume, and when I open its mildewed pages to "Verses on Kingsley's Water Babies" I can recall the wondrous sense of possibility that held me spellbound for so many childhood evenings. Perhaps my mother, who never learned to swim and who has had a lifelong terror of open water, invested her reading of this tale of an amphibious baby with a note of fear that seized my attention. And the story — with its naked water fairies, its obsessive note of fast-moving streams and sluicing tides, its protagonist's tendency, upon misbehaving, to break out in highly suggestive "prickles" — had an unmistakable erotic timbre. It's not surprising that over the years scholars have viewed *The Water Babies* as a parable of sexual awakening. Critics have described it as everything from a cautionary tale about masturbation to a wild fantasy of infantile regression in which the water itself is a symbol of the lost comforts of the womb.

I was certainly not immune to the imagery of *The Water Babies*. The story disturbed me with its hints of death and altered states and with its insistence on some vague but powerful desire that I could as yet only dimly perceive. For whatever reason, it got hold of me. It seemed to me, at the age of three or four, that it really was possible to slip, unobstructed, from one dimension to another.

The underwater world was magically accessible to me then, and I suppose I have never quite gotten over the disappointment that it did not remain so. When I was older I liked to arrive at the neighborhood pool early in the morning, before any other swimmers had had a chance to rile the surface. Standing on the edge, savoring the chlorine fumes, I would follow with my eyes the black tile track of the lane markers as they descended the concrete slope that led to the deep-water drain. The water had a harsh, denatured brilliance, and I could see every dimple of peeling paint, every lost penny on the bottom with unnatural clarity, as if I were looking through a microscope. Curling my toes over the brick edgework of the pool, I would try hard to execute an elegant dive, wanting my body to pass with barely a whisper into the untouched water.

It seemed a cruel whim of nature that as soon as I entered this world all the marvelous visual detail would disappear. My unprotected eyes saw everything through a gauzy film, and the environment that a moment before had seemed so limitless now was muffled and contracted. Even so, it was enough to be underwater, to be in another sort of place entirely. I would swim open-eyed along the pool bottom until my eyes were so stung and swollen by chlorine that at night I would lie in bed unable to sleep because of the pain.

The first face mask I owned was a dangerous toy bought at a drugstore. It was designed to fit not only over the eyes and nose but over the mouth as well, creating an air chamber supplied by two long snorkels that protruded from the top of the mask and curved upward like the horns of a goat. At the ends of the snorkels were little cages in which floated

Ping-Pong balls that were meant to stopper the breathing tubes when the diver was submerged. I had never looked through a mask before, and did not really understand what its purpose was. My hopeful assumption was that it would allow me somehow to breathe underwater, but I was not prepared for the astonishing discovery I made the first time I put it on and slipped beneath the surface. It took only a few seconds for water to leak in through the imperfectly sealed edges of the mask and flood the air space, but in that time I felt like a blind man whose vision had been restored. The human eye, as I had already discovered, is a faulty instrument when submerged. We see clearly on the surface because the fluid inside our corneas is dense enough to exert a pull on airborne light rays, bending them into focus on the retina. When we are underwater, the light rays are already traveling through a dense medium, and the cornea's power to direct them is much diminished.

When air is trapped in front of the eyes, however, the situation is more or less corrected, and the images grow sharp again (though they often appear magnified by as much as a third). When I put on that mask I did not stop to ponder the physical laws that brought everything into such supernal focus, I simply accepted these new conditions as a kind of gift. I put my wrinkled fingertips in front of the glass; they had the eerie clarity of a three-D contour map. The once-blurry images were now sharp and immediate and somehow full of purpose. My eyes were so saturated with vibrant detail that I felt like one of the saints we kept reading about in Catholic school — the unbeliever who finds himself suddenly trembling with ecstasy as he is "vouchsafed" a vision.

I don't mean to say that that first descent into the pool wearing my leaky drugstore mask was a religious experience. But years of conditioning had led me to anticipate epiphanies, apparitions, and episodes of blinding revelation. My young mind was full of ponderous and menacing spiritual thoughts. I expected at every moment a shimmering visitation from the Virgin Mary; I worried that the scrabbling sounds I heard from within the walls of my house at night were not mice or squirrels but legions of trapped souls, presided over by Satan. Catholicism enclosed me like a suffocating blanket. It ruled me and tortured me, and in time I would run away from it like a child in a fairy tale escaping from a witch's house. All the same I needed it; I craved the view it offered of something slow and eternal and unbothered, of some mystical communion just beyond the horizon.

There was another reality somewhere, to which I belonged. I knew it as soon as I looked through that mask. I was intrigued and unsettled in a way we can only be during those few childhood years when it is possible to glimpse a new world without having guessed at its existence beforehand. It *was* a new world, and simply knowing that it existed, that I could enter it, filled me with a vague contentment. I had found the wormhole — the rent in the fabric of normal existence — through which it was possible to enter some deeply satisfying other universe.

I thought of that knowledge as my secret, though of course millions of others were just as entranced, just as eager to pass through this mysterious portal. In the late 1950s, "skin diving," as it was called, was only beginning to be perceived in the light of mainstream sanity. Until Jacques Cousteau and Emile Gagnan invented the demand regulator in 1943, recreational diving had been a cult activity practiced mostly on the Mediterranean coast by men who referred to themselves as "gogglers" or "underwater hunters" and who banded together in associations with wonderful names like Club Alpin Sous-Marin (the Underwater Mountain Climbers' Club). Using only lung power, wearing motorcycle goggles and carrying tridents and spears fashioned from umbrella ribs, they plunged into unexplored coral gardens still haunted by primeval splendor, where the sluggish *mérou* and other prey fish had not yet been imprinted with the fear of man. The introduction of SCUBA — not yet a word but an acronym for Self-Contained Underwater Breathing Apparatus — made diving a more intrusive sport. All at once it was possible for human beings to linger underwater.

Scuba diving, from the beginning, had an air of dangerous allure. Every landlocked schoolboy knew of its intriguing hazards: the bends, which caused a diver's veins to fizz with carbonated blood until he died a ghastly, percolating death; and rapture of the deep, which took away his reason, filled his heart with false contentment, and drew him down into the ocean gloom. Like millions of my contemporaries, I was transfixed by "Sea Hunt," the TV series that featured Lloyd Bridges as a former navy diver named Mike Nelson. In episode after episode, Mike Nelson would be locked in deadly underwater combat with some evil agent or saboteur. Knives drawn, the two antagonists cartwheeled slowly through the water, each trying to sever his opponent's air hose and send him gasping to the surface.

Nowadays scuba diving is a rather contemplative leisure-time activity, but back in the "Sea Hunt" days it was just another test of manly worth. I learned to dive when I was fourteen, in a YMCA pool in Corpus Christi, Texas, and like everyone else in the class I imagined myself upon graduation patrolling the blue waters of the Gulf, a spear gun in my hand and an underwater bowie knife in a plastic sheath strapped to my calf. The class itself, appropriately for these martial fantasies, was run like a boot camp. Our first task was to tread water for thirty minutes without using our hands while the instructors made sarcastic comments from the side of the pool. The tanks we used were bare gray cylinders held onto our shoulders by canvas webbing that left deep impressions in the skin. The air in the tanks was delivered to our mouths through old-fashioned double-hose regulators — the kind used by Mike Nelson himself — and the long, accordion-pleated hoses fanned out from our faces like the gills of a salamander. In the classroom lectures, terse as football skull sessions, we struggled to solve incomprehensible decompression problems and watched as the blackboard filled up with physical theorems and crude sketches of ruptured lungs.

I felt as if we were training to be not merely recreational divers but members of some elite underwater commando unit. I gloried in that illusion. My diving knife, for instance, was not just a mundane tool to free myself from fishing line and other underwater entanglements, it was the weapon with which I would one day rip open the hide of an attacking shark. But beyond all the warlike daydreaming and posturing there was a deeper thrill. In the first few sessions I had trouble getting to the bottom of the pool, since none of the techniques for equalizing pressure on my Eustachian tubes seemed to work, and I was beset with constant pain in my ears. Added to that was the simple problem of strangeness — the ungainly equipment, the dulled sensory awareness, the panicky sound of my own breathing as I drew and expelled the dry bottled air. Once I had passed through all these barriers, however, I felt serene. Hanging limply on the bottom with my fins barely grazing the concrete, looking out through the glass of my face mask, whose reversed letters assured me I was protected by a "tempered lens," I felt a disembodied contentment, the contentment a soul is said to feel when it rises from the chrysalis of a cast-off body.

At the age of fourteen, my body was practically new, but I was already a little weary of its predictable sensations and its burgeoning adult demands. I was a sluggish, stolid, ungainly kid with no athletic aptitude. On my second-string high school football team, I was a nose guard, condemned to a career of thankless servitude at the line of scrimmage. But underwater my body seemed to have new properties; it had, for the first time, a grace of movement. These sustained jaunts beneath the surface carrying a portable air supply were a violation of the laws of nature, yet I felt more in conformance with the natural world than I ever had before.

Nowadays all scuba classes end with a check-out dive in open water, but in 1962 there was no such requirement. By the time I was through with my instruction I had a joyless familiarity with the U.S. Navy decompression tables and a reasonable confidence that I could handle any diving emergency that might arise in a swimming pool. Answering an ad in the paper, I bought a used tank and regulator for twenty-five dollars. The seller threw in two spear guns and a Hawaiian sling. I took the equipment home and gazed at it wistfully, but something kept me from gathering it together and heading out into the Gulf with the fish hunters who had taught me to dive. Looking back, I realize I was simply afraid. The Gulf was vast and often rough, and the offshore oil platforms where everyone dived were patrolled by hammerhead sharks and thousand-pound groupers that, according to legend, had actually gulped divers into their mouths. The Gulf of Mexico was not the point of entry I had imagined for myself — not the quiet little brook of *The Water Babies* but a roiling, dark blue mass that could envelop an intruder like a vicious storm.

My second-hand equipment, unused, was passed off to another eager buyer when I went to college in Austin. In a landlocked university town during the late sixties, when almost every aspect of life was caught up in urgent historical rumblings, my preoccupation with diving was merely a quaint relic. The reality around me was phantasmagorical enough. The ordered, limited world I had grown up in was suddenly capable of shape-shifting revelations. I remember the hysterical joy I experienced the only time I took LSD — joy because I felt confirmed in my belief that there was *more,* that human awareness did not necessarily end inside the cold gray walls that marked the boundaries of our conventional perceptions.

But strangeness has a short shelf life. Before long it turns into just another stale component of reality. As I passed through my twenties, as one daydream after another lost its conviction, I still remembered the otherworldly sensations of diving, the noiseless sauntering with which I had once moved across the bottom of that YMCA pool.

Finally I was drawn back into diving. I had long since forgotten how to work the decompression tables, and my certification card had expired, but I brushed up with a private instructor in the pool of an apartment complex and was soon reaccredited. I signed up for a three-day diving trip. The boat left from the coastal town of Freeport, and ran all night to a deep, isolated reef — the northernmost coral reef in the Western Hemisphere — known as the Flower Gardens. I hardly slept at all that night. Fretful and nauseous, I lay in my narrow bunk reading *Absalom, Absalom!* as the boat plowed through the dark, empty Gulf. The rest of the divers were older than I — physicians and lawyers who kept inspecting their equipment, their foam-lined suitcases filled with underwater cameras and strobes, with scholarly absorption. My own out-of-date equipment was stashed in an old laundry bag, and I could detect the discomfort my shipmates felt toward me. My lack of experience, along with the lack of respect for diving itself that my pitiful gear somehow implied, was a painful distraction for them.

The divemaster was a gruff, hawk-faced man who barked out the usual admonitions ("Plan your dive and dive your plan!") as we suited up on the deck after breakfast. The sea was not calm. The swells were four or five feet high, and whenever a wave crest passed beneath the hull, the wooden diving platform on the stern of the boat would plunge violently down into the trough. I had just finished attaching the second stage of my regulator to the valve of a scuba tank when I looked up to see a big sea turtle surfacing twenty yards away. The turtle's head was blunt, and its features conveyed an impression of morose curiosity. All around the creature, radiating from it, was the infinite blankness of the ocean. It was eerie and exhilarating to imagine the life that turtle led, as solitary as a comet wandering through space.

The divemaster took me on as his buddy, and I watched from the boat as he disappeared beneath the swells and followed the bright yellow descent line to the bottom. I took a few practice breaths (hawooo-*huhh,* hawooo, *huhh*), waited until the pitching boat was close to the surface of the water, and jumped in with a wide, flat-footed straddle. The sensation of jumping into the open sea that first time was as startling and absolute as I'd always imagined the sensation of sky diving would be. All at once I was alone in the firmament, and though I was not hurtling downward, the feeling of suspension was just as intense.

It took me three or four calculated breaths to calm myself and look down past the blunt, swaying tips of my fins. The water was a deep blue, and against this backdrop the expanding bubbles that arose from the divers below me were a brilliant silver, so sharply defined they looked like solid metal disks hurtling toward the surface. I grabbed the descent line and lowered myself hand over hand. I had not gone five feet when my ears began to hurt. The trapped air in my Eustachian tubes felt as dense as mercury, and there was no way I could relieve the pressure. I moved my jaws up and down, I pressed the mask against my face and exhaled, I swallowed and rocked my head from side to side, but the pain just grew more concentrated. I must have stayed there for ten minutes until finally, bit by bit, the pain lessened and I was able to sink slowly to the coral bank.

The divemaster took my hand and led me around, pointing out fan worms and Christmas tree worms, which would pop back into their burrows as we approached them; the mustardy growths of stinging fire coral; a crevice from which a small spotted moray eel protruded, its flat body swaying in the current like a banner. He let go of my hand and gestured with a wide sweep of his arm at the seascape before me — the bulbous mounds of brain coral, the fissures and shallow canyons floored with blinding white sand. It was a theatrical, half-joking gesture, but I chose to read it seriously. Here is the place, the divemaster's outstretched arm seemed to indicate, that you have been seeking.

But I had only a glimpse of it on that dive, since I had depleted most of the air in my tank during my slow and stressful descent. I did not understand the radiant and unsettling forms of life that stretched out before me — the corals and fishes and anemones and the specks of plankton that swept over the reef like motes swirling in a beam of light. There was no way to focus on any one piece of it, to find some crucial illuminating detail that would help me perceive its purpose. The waving tentacles, the darting fish, the ceaseless secret business of the reef filled me less with wonder than with anxiety. My desire to comprehend this place was a kind of panic.

As I gained altitude, rising toward the surface and the shadow of the boat seventy feet overhead, the separate components of the Flower Gardens became a compacted, colorless mass. Barracuda hovered along the descent line, their teeth exposed in a carnivore's rictus, their eyes ticking off my passage. Behind them the ocean backdrop had the deep blue tincture of an approaching thunderhead. I wafted to the surface like a figure in a dream, and when I climbed back onto the boat the memory of that first brief visit to the reef taunted me until it was time to go down again.

That was years ago. The dive is recorded in my first logbook, a pale blue vinyl notebook whose waterproof pages, bound with a plastic spiral, provide space for indicating visibility, bottom time, maximum depth, and temperature. I soberly entered this data after the dive, and on the lines reserved for "Comments" I merely made a few guarded observations: "shallow coral heads; thermocline at 80′; barracuda; strong surface current." There was no room to say more, and anyway I loved my terse declarative observations, with their authoritative semicolons. At the time those words seemed as well chosen and concentrated with meaning as the words of a haiku.

Eventually that first logbook was filled up, and then another. Brusque as they are, they seem to me now like dream journals, like diaries kept in a fever. Reading back through them fills me with nervous energy and with a kind of resentment that they do not constitute, as I had once hoped they would, the autobiography of an underwater pilgrim. The entries are too spotty for that, though the memories they evoke are vivid. On one page I am groping blindly along the mud bottom of Matagorda Bay, searching with a group of Texas archeologists for any nails or buckles or harquebus parts that might indicate the wreckage of the flagship of the Sieur de La Salle. Overhead the engine of the research boat is running, its prop wash deflected downward by means of an elbow-shaped aluminum housing that blasts away the oozy overburden, bathing us in turbulence and total darkness.

On another page I'm diving among giant clams off an island on the northwest coast of Madagascar. The island's main village is called Hellville. Its forests are populated with bug-eyed lemurs, its beach patrolled by a giant aldebara tortoise named Caroline who presents her neck, thick as a python, to be stroked. Underwater, I pick up an empty cowrie shell from the sand bed and run my fingernail along its serrated aperture, producing a sound like a finger raking the teeth of a comb. Staring at the blurry brown spots on the back of the shell, I suddenly flinch, contracting the muscles at the back of my neck with the involuntary wisdom of a prey animal alert to its most vulnerable points of attack. The gleaming hide of a large shark appears from above in my field of vision. I watch it sailing ominously over the reef, which with its giant clams and undercut mushroom growths of coral looks as placid and picturesque as a Hobbit village. The shark's teeth are exposed in its lipless mouth, its nose tapered to an artful wedge. The creature's form is beautiful, but the utter blankness in its eyes is terrifying.

Or I am diving off the coast of southern California, at a place called Begg Rock, a barren islet that barely breaks the surface of the Pacific chop. I'm wearing a full wetsuit. The only place the water touches my skin is in the gap between the top of my mask and the rim of my neoprene hood, so I feel as if I'm swimming with an ice cube on my forehead. Underwater, the rock is covered with scallops and anemones, their tendrils and tentacles — bright orange or red or yellow — vivid against the grim asteroid shading of the rock itself. Into view comes a face so familiar and unthreatening that for an instant I almost raise my hand to wave. It is a sea lion, soaring toward me with languid winglike strokes of its pectoral fins. The sea lion comes to within three feet of me, studies me frankly with its moist, doglike eyes, then veers away, its body spinning like a bullet. I watch it go, watch the dark blue water enfold it, and am left with an odd thought: the sea lion is now swimming through the Pacific with my image lodged in the circuitry of its brain.

The logbooks filled up, but the gaps between dives became longer and longer. I had three children; I had spiraling responsibilities of the sort that forced me to regard diving as an expensive recreational activity rather than as a life's mission. I looked at those logbooks, at the paltry entries in the "total bottom time" column and felt almost ashamed, as if the few underwater hours totaled there were the pitiful record of an abandoned spiritual quest. I read through the diving magazines — the ads for resorts and live-aboards in Aruba and Bonaire, the models posed in hot pink wetsuits, lasciviously displaying some new gauge or contour-fitting buoyancy compensator — and grew self-righteous and edgy. I could not tolerate the notion that diving might be merely a hobby, but at the same time I envied the people in those magazine layouts. I envied their seemingly eternal leisure time, their perfect bodies, their gorgeous equipment, their casual, uncomplicated appreciation of a world that I had hungered for all my life but that continually eluded me.

I sat on the edge of the bathtub as my youngest daughter discovered she could put her face underwater and open her eyes without coming to harm. Six inches of water, and nothing to see except the white curving shape of the tub — yet night after night she plunged below the surface for as long as she could hold her breath. I would watch her arms and legs thrashing as she tried to surge down into the depths she imagined were there. I knew what she wanted. She wanted to be at large in this sudden new territory, to pursue the illusion that she was fit to inhabit it, that its strangeness would embrace her.

I had been there when her two older sisters had made the same discovery, and I remembered that same look of pride on their faces when they broke the surface — the beaming look of voyagers returning triumphant from the abyss. Sitting on the edge of the tub, a towel ready on my knee, an encouraging smile on my face, I felt my heart shrink with envy and then, a moment later, swell with resolve. Somehow or other I would get this out of my system. I would go to some island and plant myself there for a period of months, diving every day along the same reef. To make this scheme respectable I resolved to teach myself about fish and coral, about bristle worms and nudibranchs. I could truthfully maintain that I was traveling to the Caribbean to study the natural history of a coral reef, but the deeper motivation was not as clear or, perhaps, as worthy. I wanted to be, at least for a time, my underwater self.

2

Dry Dock

The island I chose was Grand Turk. Eight miles long and two miles wide, it is one of a series of tiny islands clustered north of the Windward Passage, which separates Cuba and Hispaniola. The islands themselves form a scattered little nation known as the Turks and Caicos, a British Commonwealth colony long since gone to seed and now fitfully trying to revive itself through tourism and offshore banking.

I chose it on a whim, because I hadn't heard of it before and it sounded lost to the world, a stern place with no nightlife and no Club Med, where you could imagine that the ocean still prevailed, beating down every trivial notion before it arose, ruling every thought. A coral wall, bountiful and largely untouched, ran along the length of the island just off the west shore and plunged down into an 8,000-foot ocean chasm known as Turks Passage. People said that sometimes when you were diving on the wall you could hear the distant deep-water ping of Russian submarines as they cruised through the channel. And in the winter humpback whales passed through on the way to their breeding grounds on the nearby Mouchoir and Silver banks. Sometimes, if you were extremely lucky, you would see them moving along at the edge of sight, filling the canyons and coral crevices with their melodic lowing.

I went to Grand Turk because I had heard about a diver who once drifted out from behind a large coral head and happened to see what no human could ever reasonably hope to see: a humpback whale in the act of giving birth.

The island was treeless and brush-covered, with raw cliffs of pockety limestone rising from its northern end like the prow of a ship. Once it had been wooded, but the salt merchants of centuries past had cut down all the trees, wanting instead the unbroken evaporative power of the sun. The salt ponds, known as salinas, still dominated the middle of the island, though they had not been worked for decades. Bits and pieces of arcane machinery still stood rusting on the edges of the salinas, and egrets and black-necked stilts waded near the idle boiling pots that had once channeled water in from the sea. The bars in town were filled with blokes and mates and mons, with retirement-age swashbucklers and goldbugs and men who, when asked their profession, would say, "Oh, I have some — uh — investments." The few roads were dominated by wild donkeys, by cattle with missing or ingrown horns, by hermit crabs crawling across the asphalt in their sun-bleached shells. The iguanas, I was told, had all been eaten by cats.

I arrived at Grand Turk on a Sunday. On the flight from Miami I sat next to a man who had boarded the plane holding a fragrant box from Pizza Hut in one hand and a ferret in the other. The man's name was Roberto; the ferret's name, he said, was Digger. Roberto was in his late thirties. He had olive skin, a fastidious mustache, and elegantly receding hair. His clothes were studiously casual. He told me he had been born in the DR — the Dominican Republic — but had lived in Grand Turk most of his life. His present profession was reinsurance.

Digger was as frenetic as a hamster. She wriggled in and out of Roberto's lap and, nostrils wide open, ran her snout along the seams of the pizza box. There was something disquieting about her movements. Watching her liquid, weasely squirming, I began to think about motion sickness. Every so often she would pull up and regard me with a look of alarmed curiosity. She appeared desperate for any sort of information.

"She wasn't eating," Roberto said when I asked how she happened to be on the plane. "I was worried about her, so I took her to Miami. The vet examined her very carefully but could find nothing wrong. Finally he asked me, 'Your apartment on Grand Turk. Is it air conditioned?' I said that it was. He asked if I kept the air conditioning on all the time, and I told him that I did. 'That is the answer!' he said. 'Your ferret, my friend, has been hibernating!' "

He held Digger by her hindquarters as she strained forward and put her head into the seat pocket. Watching her ceaseless, supple movements, I was struck by how graceful and efficient she would seem underwater — if, like her otter cousins, she could swim. She reminded me of that sea lion off Begg Rock; she had the same searching eyes, the same fluid body.

The Grand Turk airport was a cinder-block building imperfectly shaded by a few casuarina trees. It was early October, off-season, and the sun was still strong enough to make my skin contract as I stepped out of the air-conditioned cabin. The little airport was jammed for the arrival of the twice-weekly flight from Miami. There was the usual crowd of baggage handlers, taxi drivers, and petty entrepreneurs, as well as a few policemen in immaculate colonial uniforms and a ruddy-faced gentleman fanning himself with a planter's hat, whom I correctly assumed to be the governor of the Turks and Caicos. Homeward-bound divers — in T-shirts bearing florid likenesses of dolphins or manta rays or the simple logo of a diving equipment manufacturer — stood about sipping beer and waiting to board. The hair on their forearms was bleached white, and the skin on the back of their necks was peeling off in sheets as crinkly as parchment. The divers appeared at ease, played-out, sated. They may have been wage slaves back home, but here in Grand Turk they had the languorous bearing of aristocrats.

My luggage and dive gear failed to emerge from the plane, and the baggage inspector was unexpectedly sympathetic as he sorted through the oddball contents of my carry-on bag. There was an empty ledger book ready to be filled with underwater "data" that I planned to enter in a crabbed scientific hand. There were field guides to fish and coral, a one-volume *Encyclopedia of Aquatic Life,* books by William Beebe and Jacques Cousteau and a copy of the collected poems of Wallace Stevens that I had grabbed off my bookshelf at the last minute, not knowing why. In a side pocket were three thick carpenter's pencils, each with a neatly drilled hole for attaching the pencil by string to an underwater slate.

"You are a biologist?" the inspector asked.

"No," I said, "a tourist."

I filled out a form for my lost luggage and walked outside, feeling the slightly morbid unease of the solitary traveler. I could not see the ocean, but I could sense its luminous, taunting presence. The governor climbed into his Austin-

Healey and was driven away. I shared a taxi with a group of Canadians headed for the *Sea Dancer,* a ninety-two-foot dive boat that is headquartered in Grand Turk and sails the waters of the Turks and Caicos for a week at a time.

A cloud of white dust followed the taxi along the island roads. The ground, parched and austere, reminded me of the bolsons and caliche flats of desert Texas. Along the way to the dock we passed groups of people walking toward church beneath the burden of the sun, the women in faded print dresses and white gloves, the men in dusty black suits. Young men galloped horses bareback down the main streets, and children wobbled along on rusted-out bikes, painfully working the exposed pedal cranks with their bare feet.

Beyond the colonial rooftops of Cockburn Town, the blue ocean rose into sight. Its cartoon brightness was intoxicating, and it seemed splendidly unnatural, like the imagined sky over some planet at the edge of the solar system, full of glowing phenomena and rapturous colors that human eyes had never seen. The taxi pulled up onto the wooden dock and stopped next to the *Sea Dancer.* The crew, in khaki uniforms and sun-bleached beards, relieved the arriving passengers of their baggage and offered them soft drinks from a tub of glistening ice. While the driver settled up with the Canadians, I stood at the edge of the dock and looked down into the water. The bottom was at least forty feet down, but thanks to the pure marine optics — the way the sunlight flashed on the mirrorlike brilliance of the white sand — it looked much closer. Here and there in the turquoise water were patches of dark blue, indistinct as cloud shadows. These were coral heads, and from the surface I could see fish swarming about them. The coral was spotty here, but a few hundred yards out it began to gather into a compacted mass, and the ocean abruptly changed hue, no longer aquamarine but an abyssal shade of blue. That precise, unwavering line was the wall, dropping to its 8,000-foot chasm.

Bathed in the sight of that water, I felt what I perceived to be a sexual longing. I wanted to remove my clothes and dive naked off the dock, never to surface again but never to die. When I was very young I had experienced erotic childhood dreams of drowning. I would be lying on a beach, and the tide would come in, moving slowly up and over my body like a rising bath. The water would creep up to my mouth and nose and finally, to my excitement, cover them. I would hold my breath as long as I could until I was overcome by the knowledge that there would be no more air. This was an outrageous, unbelievable fact, and in my dream I struggled against it with authentic desperation, yet part of me waited patiently for something better than oxygen, something beyond the turmoil that was not death but a wonderful transformation that would send me like a flooding tide to every unknown corner of the universe.

It was lust that I felt there on the dock — a cratering, yearning emptiness, a fear of never being satisfied. And yet there it was, fulfillment itself: the clear blue ocean, blue as ice, the new country to which I had come to claim my rightful citizenship.

I found the Island Reef Hotel on the other side of Grand Turk, the windward side, where the island rose up forty or fifty feet from sea level and formed a spiny ridge. Just below the ridge, fronting the ocean, sat the hotel. It was an anonymous structure, a line of one-story apartment units with an open-air bar and restaurant at the far end. Beyond it lay a sheltered lagoon, a big shallow sand flat dotted with patch reefs that extended for perhaps a mile. Unlike the underwater topography on the other side of the island, which changed abruptly several hundred yards offshore at the precipice of the wall, the reef structure here was gradual, ending in a jagged and ill-defined front, and the deep water was distant. Looking out from the top of the ridge I could see a thin line of white water marking the reef crest, the turbulent zone where the wind-borne water of the open sea crashed against exposed brambles of elkhorn coral. From the shore to the reef crest the lagoon was a blue savanna that seemed to derive its brilliance not from the sun but from some powerful light source deep beneath the surface.

My luggage, with all my diving equipment, was not scheduled to arrive for another three days, and looking out at the water I was heartsick with frustration. I walked into town in the heat of the afternoon, following the paved road down the ridge. There was brush on either side, dense chaparral in which I could hear small lizards and hermit crabs rustling about. I noticed one or two of the Turk's-head cacti that had given the island chain its name, squat barrel-shaped plants topped with a red fez. Wild donkeys — a mother and colt — watched me approach, then let out a tortured bray and moved off in a rattletrap gait. Farther along the road I encountered a horse trotting placidly in the opposite direction with an empty ice cream carton in its teeth.

Drenched in sweat, I arrived at Front Street, a decrepit waterfront promenade that forms the main street of Cockburn Town. At this hour of day the traffic was light. A policeman, his uniform miraculously crisp in the damp heat, walked pridefully down the rutted street, and after him came a man driving a donkey cart that lurched and swayed over the potholes like a prairie schooner. From a warped wooden pier that extended from the sea wall, a group of boys fished with handlines. I noticed a few undistinguished buildings of recent vintage, mostly housing branch banks and understocked hardware stores, but the rest of the architecture harked back to the days when the first Bermudian settlers arrived and built cool louvered houses with wide verandahs facing the sea. Front Street was lined with such evocative old structures — with white-washed houses set back on an acre or two of desiccated lawn, with ancient rainwater cisterns and carriage houses and bougainvillea flourishing over crumbling stone fences.

I walked up and down Front Street for an hour, keeping an eye on the deep blue reef line that marked the wall. In a few days my gear would arrive and I would be soaring along the face of that wall the way an eagle soars along a cliff face, its mind and body in perfect conformance with the elements, not a muscle, not a thought out of place. But for now I had to sit tight. I stopped for an early dinner at a restaurant overlooking the water, and as I waited for my conch chowder I watched the sea's calm surface, entranced by the small, perfect rollers that slipped onto shore and died without a whisper on the beach. The little waves traveled over the water like rolling pins over a sheet of pie crust, ironing it down smoother and smoother.

As the sun began to go down, the wind came up, a steady breeze from the south that broke the pattern of the waves and created a sloppy, directionless chop. A kitten darted out from under a nearby table and sunk its claws into my bare leg.

"Is he bothering you?" the waiter asked as I pried the kitten loose. One of his claws was stuck deep in my flesh as tenaciously as a fish hook.

"No," I said, "not at all."

A motorcycle roared to a stop outside and Roberto came in, accompanied by a young woman with dark hair and a sleepy, exasperated look.

"How's Digger?" I asked.

"Sleeping," he said, accepting my invitation to share the table. "But I've set the thermostat higher this time."

He introduced me to his date. Her name was Antonia. She was from the DR and spoke no English. Roberto took out two cigarettes, lit them both at the same time like Paul Henreid, and passed one to Antonia. He set a six-inch lock-back knife on the table and toyed with it while he talked, pulling the blade halfway out and then letting it spring back with a snap.

While I ate my chowder Roberto talked about his boyhood in Grand Turk. He had grown up without a father. He said his childhood was "difficult."

"It used to be the Americans had two bases here. A navy base and an air force base. The Americans used to throw a lot of shit out of the mess hall. Pieces of meat and stuff. We called it *garbago,* because it was garbage. You'd bring home a pound of meat for your Sunday dinner. It had teeth marks in it, it tasted like diesel, but it was good shit. Before the Americans came, Turks Islanders had never seen meat. This was our first introduction to bologna, salami, steak. We knew nothing about electricity until fifteen years ago. We got our clothes from Church World Services. The first suit I ever saw was a Salvation Army suit.

"I was a surveyor for a while. I remember going to this little island. We took all these bulldozers and shit. And I am working around this bulldozer and guys are arguing: 'It's a Mustang. No, it's a Corvette.' They're saying, 'What a beautiful car!' and the thing is a fucking bulldozer! Another guy comes up to me, looks it over for a long time, and says 'Where's the anchor?' "

While he talked in English, Antonia looked off patiently into space. From time to time Roberto would put his hand on the back of her neck and give it a proprietary squeeze. The wind kept growing stiffer.

"In your country," Roberto was saying, "kids learn to walk first, but in this country — literally — we learn to swim first. I don't even remember learning to swim. It was just always there. I remember diving for turtles. No masks, fins, or anything. I've gone eighty feet with nothing. I'd dive anywhere there was a good reef where I could get a turtle.

"The people here live from day to day. There's very little employment, okay? We have devised a system of survival where you eat a lot of fish. I can put on my grits to simmer, go to a place I know, catch a fish and clean it and be back before the grits are done. Those of us on the island, we *know* we will not go hungry. We may be two hundred years behind the U.S., but there is that primitive peace we do not want to lose. Ah, but the sad thing is, we know we must lose it."

The wind that had come up was the leading edge of a tropical storm named Emily. Three days earlier Emily had been a hurricane; it had torn through Hispaniola, had broken up in the mountains there, and had begun to regenerate as it made its way through the Windward Passage. When I got back to the Island Reef that night, the wind was blowing at thirty miles per hour and the rain was coming from the east in lateral bursts. Though Emily never recovered enough strength to reach hurricane force again, the weather stayed bad for four more days, and the flight from Miami that would have brought my missing diving equipment was canceled.

Morose and a little homesick, I waited out the weather in my room, reading about fish and coral, trying to prepare my mind for the chaotic beauty of the reef. Reggae music ("I appreciate the hurt that you brot to me/Thanks a whole lot/

Thanks a whole lot") came drifting in from the radio in the bar. Several times a day I would wander down and play darts with a group of British civil servants who, when sufficiently drunk, would break out into a woolly version of "We Are the World."

"Do you know what it's like living here?" one of the Englishmen said to me one night as the wind howled through the window mesh. "It's like living in an unreleased Humphrey Bogart movie."

The storm passed finally, and in the wake of all that rain the island's limestone had the gleam of polished marble. From out of the clear sky the Pan Am jet descended, bringing my luggage at last. I set my dive bag on the bed and unzipped it. My wetsuit was on top. I unfolded it, fussily smoothed out the creases in its neoprene skin, and hung it in the closet. My mask with its prescription lens had arrived undamaged. I strapped it on and looked around. My vision was sharp, perhaps even a little better than with glasses, and light seeped in through the transparent rubber skirt that framed my face. I looked in the mirror and felt a little crestfallen. My face was too broad, my head too bald. The mask gathered my features into a puckered grouping, the tight strap caused the hair on the sides of my head to sprout outward in comical tufts, the snorkel flapped overhead like a thick, flexible antenna. What bothered me was not that I looked ugly, but that I looked awkward. I wanted to be a graceful creature underwater, with every hair and strap and hose as functionally in place as the feathers of a hawk.

The mask was a recent purchase, but my fins were old and heavy and out of style, like an ancient cherished pair of Sunday shoes. And beneath them was the dive knife that I had never unsheathed in anger and the white slate that still bore the cryptic traces of an underwater conversation—

"What was that?" "A wolf eel" — from a dive a year earlier in the California kelp beds. The last things I pulled out of the bag were a buoyancy vest and a regulator with its tendrils of hoses. The vest, the regulator, and the attendant gauges were all made by U.S. Divers, the company founded by Jacques Cousteau.

Years before, I had met Cousteau when I interviewed him for a magazine article in the offices of the Cousteau Society in New York. He wore a close-fitting maroon suit of some exotic cut, and at sixty-eight he was still fit in a pallid European way. He had a carelessly efficient, highly bred body — long torso, narrow shoulders, spidery limbs. Sitting behind his desk, he puffed on a cigar and answered my questions with a pronounced air of impatience. I wanted him to like me, in part because I felt real gratitude to him for the device he had helped to invent: the demand regulator, which when attached to a tank of compressed air became the Aqua-Lung. Cousteau, the man sitting across the desk, had given me and millions of others this intimate gift: the ability to breathe and roam freely beneath the surface of the water. In the manner of a journalist trying to elicit a lyrical quote, I asked him to reflect on his own emotional response to this great invention that had been accepted so heartily by the world.

"No," he snapped, dismissing the idea entirely with a wave of his cigar. "I have no feeling for that invention. To me it is dead."

Now, looking down at the diving equipment spread out on the bed, I remembered that comment. At the time it had seemed a strange and cold thing for Cousteau to say, but now I understood that it was just a weary answer to a sentimental and probably commonplace question.

Still, I didn't quite see how he could disown all of this

marvelous stuff. This array of rubber and chrome objects, these gauges and hoses with their watertight O-rings and quick-release plastic buckles, were among my most cherished possessions. They were more than "gear"; they were a costume, the magic raiment that would allow me passage to the reef.

3

Harmonium Point

There were always parrotfish on the flats. You would hear them gnawing on the coral, breaking off chunks with their huge front teeth. The teeth reminded me of chisels, or of the overgrown incisors of cartoon donkeys. The parrotfish swallowed the pieces of coral and processed them through the pharyngeal jaws in their throats. I imagined these jaws as a mighty grinding machine, its cutting edges as hard as diamonds, that pulverized the coral in powdery explosions as it came down the conveyor belt of the fish's throat. The process liberated the filigree of nutritious algae that grew on the coral. This was what the parrotfish ate. Everything else became dust, and all over the reef you could see little downward drifts of sand, the grains sometimes sparkling when the light caught them, as the parrotfish evacuated it from their brilliant bodies.

That first week on the reef I found the water still murky from the passage of tropical storm Emily, and it was laced with upwelling cold currents. But compared to the silty

Texas lakes I was used to diving in, the reef was ravishing. I could see perhaps fifty feet. We would moor the boat just above the abyss of the wall and descend thirty or forty feet to the sand flats on the landward side. The sand was as pure as a snowfield, and when the sun was high there was a mild glare. Seaward the sand disappeared beneath a solid hedge of reef crowned with sea fans and other soft corals that waved in the surge like windblown plants. On the other side of the hedge was the wall.

Usually I saved the wall for the second half of the dive, not wanting to approach it too suddenly, not wanting the sensation of soaring out over its lip to become too common. I preferred to sink to the sand beneath the shadow of the boat and spend a few meditative moments savoring the knowledge of where I was. I felt the way I had felt the one time I went to Paris: never able to trust that I was there, to convince myself that I had actually crossed the Atlantic and that the great sights before me — Notre-Dame, the Arc de Triomphe, the Winged Victory in the Louvre — were not hallucinations but things that I truly beheld, and that in some odd way beheld me in turn. It was as strange as Paris here on the reef, as strange and as magnificent. When I first slipped off the diving boat and sank to the sand I was seized with a puritan insight: it seemed to me that the surface of the island was somehow being punished with drabness for the profligate beauty it displayed beneath the surface.

The flats stretched out before me like a miniature Sahara, the sand laced with snail tracks that looked from on high like the tracks of desert caravans. Coral oases loomed in the distance with the hazy, teasing presence of a mirage. Some coral heads were no larger than floral arrangements, others were massive entities, worlds unto themselves across whose orbits speeding fish traveled in meteoric swipes. The coral heads themselves were full of crevices and overhangs through which other sorts of fish wriggled in and out, back and forth, searching for food or refuge. I knew the names, at least, of most of them: bluehead wrasses, rock hinds, blue tang, groupers, a dozen different kinds of gobies darting about in their start-and-stop fashion above the fissures and buds of the coral surfaces. Doctorfish, trunkfish, cowfish, boxfish, butterflyfish, lizardfish, hogfish, squirrelfish, parrotfish — their names reinforced the perception that always took hold of me underwater, that I was in a kind of counterworld, where every feature and principle of surface life had its alien obverse. Including me, the manfish.

But it was the parrotfish that fascinated me the most, perhaps because they were so hard to miss as they swarmed over the coral, rasping and biting and grinding it to dust. There were eight or ten different species. Some were small and as dull in their coloring as bass. Others — the rainbow parrotfish, the midnight, the stoplight, the blue — were so glorious that each scale covering their skin was like a panel in a stained glass window, the color always ebbing or flaring as the light changed sixty feet above.

Parrotfish are sequential hermaphrodites. As they grow older, they sometimes change sex. A fish that begins as a female may end life as not just a male but a supermale, a massive, radiant creature that is soon to die but seems triumphant in its terminal masculinity.

I saw supermale parrotfish often on the flats. The blues were the hardest to miss. They were three feet long, broad and weighty, with a royal blue coloring that the light brought out in ever more intoxicating, ever more complex hues. When they were younger, their heads had sloped to a wedge, but as supermales they had the bulbous foreheads of sperm whales. They moved slowly over the sand, not in a hurry, their every move announcing power and experience. Unlike the more junior parrotfish, which swarmed over the reef by the hundreds but were essentially solitary, the supermales usually traveled in groups of three or four. I watched as a group hovered above the sand, seeming to confer and deliberate among themselves until one of them trimmed his fins and sunk to the bottom like a submarine taking on ballast, plucking some calcareous morsel from the sand with his teeth.

I would watch the supermales until they dissolved into the haze, and then I might turn my attention to the garden eels that sprouted from the bare sand like a sparse, swaying mat of grass. Sometimes there would be what seemed like an acre of them, a great field of miniature eels protruding halfway out of their burrows, heads turned toward the current and bodies moving with rhythmic undulations that put me in mind of a thousand cobras rearing from a thousand baskets. I could get within five or six feet of them, but then the eels closest to me would begin to disappear as if some invisible scythe were preceding me and cutting them down. They popped back underground with startling speed, and the sand closed over the holes of their burrows, leaving no evidence that they had ever been there.

When I was feeling patient I would sink to the bottom and creep up to them inch by inch, exhaling quietly and steadily until I was close enough to see the silvery glint of their eyes, the only feature I could discern on those waving stalks. In the next moment, as I advanced weightlessly on the tips of my fingers, they would be gone. The way they withheld themselves from my sight pestered me and struck me as slightly malevolent. They made me think of a television program I had watched when I was very young, a puppet show called "Time for Beany," about a little boy and a sea monster. The monster's name was Cecil, and the viewer saw only his top half. I was too young to understand that he was a puppet, that his bottom half was only a hand inserted into a sock. *Something* — the rest of Cecil — was below the edge of the screen, and it aggravated me that I could not see it. Standing in front of the television, I would set my eyes against the glass at the top of the screen and peer down, hoping through this angle of sight to catch a glimpse of Cecil's mysterious extremities.

The garden eels had that same disturbing, unrevealed quality. In my fish book back at the Island Reef was a cross-section diagram of an eel in its burrow, looking like a wood screw embedded deep in a two by four. The burrows are reinforced by the mucus the eels secrete from their skin. Though the creatures have the ability to swim, and the males sometimes venture out in search of better mating opportunities, in general they are as rooted to their spots in the sand as if they were potted plants. They mate by twining around each other and releasing eggs and sperm, the ends of their bodies still touching base in their burrows.

Usually I lingered behind the other divers, respirating on the bottom with the eels while the rest of the group headed straight for the wall. I had hooked up with a group called Blue Water Divers, which at that time was one of the two diving operations on the island. Blue Water's sole significant asset was a twenty-three-foot hydrofoil boat with a sixty-five-horsepower outboard that could zip out to the reef in only a few minutes. Underwater, I was always aware of the boat hovering above us like a protective cloud and of Mitch Rolling swimming watchfully along on his back, his arms folded across his chest in an attitude of meditation.

Mitch was one of the two owners of Blue Water. In the time I was on Grand Turk I never met his partner, who had gone back to the States on urgent family business and lingered. He and Mitch came from Ames, Iowa (the parent company of Blue Water was called Iowa Undersea Adventures). Upon graduating from high school eight years earlier, they had lit out without hesitation for the tropics.

"We had this dream ever since we were sophomores," Mitch told me. "We were going to go to an island. For the last two years of high school we researched where to go. We wrote a letter to the Tahiti immigration office, but when they sent us back an application in French we blew it off. There was this pilot in Iowa, though. He'd done a little charter work in the Bahamas, and he asked us if we'd ever heard of the Turks and Caicos. We hadn't, of course, but we went to the travel agent the next day in Ames and she looked it up. She found out that Air Florida flew there from Miami for two hundred and fifty dollars round trip. That was the key right there."

Mitch had originally come to the island with the intention of finding a congenial place to play his guitar, and over the years he had emerged as Grand Turk's only musical celebrity. My second night on the island, I had gone to see him in the thatch-roofed bar of the Salt Raker Inn, where the tables were strewn with laminated fish charts and back issues of *Skin Diver*. He sang standards by Jimmy Buffett and Kenny Loggins and, upon request, the theme from "Gilligan's Island."

Mitch lived and worked in a house just north of the center

of town, thirty yards from the sea wall. The mummified head of a jewfish, as big as a boulder, sat on the front porch. Its jaws hung open, and in the middle of its head was a hole where it had been speared by the man Mitch and his partner bought the dive business from. The living room of the house resembled the cabin of a ship, with dark wooden beams and white walls covered with nautical charts. In a brick-and-board bookcase sat a dusty set of Great Books, and above them a collection of antique bottles, shells, barracuda skulls, and the armored headplates of spiny lobsters. From the ceiling, running the length of the room, hung a life-size great white shark, sculpted in papier-mâché.

I would ride up to Mitch's house every day at eight o'clock on my rented scooter, my mesh bag of dive gear balanced on my lap. Mitch would be in the compressor shed behind the house, filling the day's tanks, singing "Island Girl" or some other song I recognized from the cassette he offered for sale on the Manta Ray label. The tanks were aluminum, taller and lighter than steel but still heavy when filled with compressed air. I carried them two at a time across the street, holding them awkwardly by the valves, picking my way across the rocky ground in my flip-flops. When I had lined them up on the sea wall, ready for loading onto the boat, they made a pretty sight, a row of bright yellow cylinders against the glorious blue of the ocean.

Usually five or six divers had signed up for the trip. They might be solitary accountants or civil servants who had come down to Grand Turk on business, or honeymooning couples who had come over for the day from the Club Med on Providenciales. Or they might be serious divers, touring the Turks and Caicos for three or four weeks at a time, carrying their gear in three-hundred-dollar backpacks. The appeal for them, and for me, was the unspoiled bounty of the reef — the lush coral, the swooping columns of fish. Grand Turk had not yet been "dove out." The place was calm and undiscovered, lacking any amenities except the reef itself. This was not yet a major dive resort, where the cattle boats disgorged forty or fifty underwater tourists at a time; where the coral was broken and bruised by the careless swatting of fins; and where the fish, ruined by handouts, swam up to the divers expecting an aerosal burst from a can of Cheez-Whiz. "Yeah, this is pretty great," a paper salesman from Seattle told me one day as we headed in from a dive. "Come back in ten years, it'll break your heart."

Mitch had made more than a thousand dives on the Grand Turk wall, and in that first week he took me to all of the main sites, some of which he had named. A few of the names were evocative: the Black Forest, for a spot beneath the overhang of the wall where fronds of black coral grew sheltered from the brunt of the sun; the Amphitheater, for a site distinguished by a vast sloping bowl of sand. Other names were flatly informational, simple declarations of underwater landmarks: the Anchor, the Canyons, the Tunnels. During my months on Grand Turk I developed an intimate acquaintance with most of these places, but on those first few dives I could scarcely tell them apart. At the beginning I saw only the repetitive textures of coral and sand, the patterns in an endless bolt of fabric that seemed to unspool below me as I drifted along. And the essential experience was constant: the loitering in the sand flats, the lazy saunter upcurrent along the top of the wall, and the trip back along its face with our bodies hanging over the abyss — an enchanted dispensation from the laws of gravity.

At some sites the wall was a sheer vertical drop into the darkness, at others the descent was interrupted by a staggered series of terraces. But there was always the feel of deep ocean beneath, a sensation like no other on earth: hanging unsupported along the face of a precipice with miles of blue nothingness ahead and the black ocean pit below. I felt like Wile E. Coyote in the "Road Runner" cartoons, when he inadvertently runs past the lip of a cliff and stands there in midair, looking sheepish as he awaits the plummet.

But the plummet never came, and that was the magic. Instead I would let myself drop, descending at my own pace, following the wall for fifty, sixty, seventy feet as the color bled away and the hard, expansive corals gave way to softer things, to deep-water lace and the looping coils of sea whips that grew straight out from the wall. It was a region of colorless filigree, of waving, pliant forms that gave no resistance to the crushing weight of the water. I would punch the inflator button on my buoyancy vest as the sea's increasing density threatened to draw me deeper, into what I imagined as a region of ghastly bioluminescent fish and steaming ocean vents. A part of me wanted to go there, but I hovered instead at a sensible depth, looking out and not down, out toward the trackless blue water beyond the reef, water that was like a magician's cloak from which anything could materialize.

One site in particular kept drawing me back. It was named the Library in honor of the shoreline landmark — a pink stucco library, formerly a church — that was visible two hundred yards away on Front Street. In one area here the wall was cut through with two deep canyons, creating a promontory that looked out commandingly over the empty ocean. Some places have a kind of power, a provocative allure, simply by virtue of the way they inhabit space, the way they sit there, as if sentient and alert to their existence. This promontory at the Library drew my eye and focused my vision along its narrowing platform until I saw what it wanted me to see: a triangular shelf of coral rock, vectoring out into the blue.

At the very edge of the shelf stood a barrel-shaped growth of cavernous star coral — *Montastrea cavernosa.* (I had been teaching myself the Latin names since the common names for coral species differ from book to book.) It was an elegant specimen — three feet high, slightly squarish, colored a muted shade of forest green that made the coral boulder look as worn and soft as an upholstered ottoman. The polyps in their bulbous cups were bunched up on the surface like biscuits in a baking tin.

The perfect location of the coral upon the promontory, its deeply satisfying presence and symmetry, made me think of Wallace Stevens, whose poems I had been reading back at the Island Reef in between entries of *The Encyclopedia of Aquatic Life.* I remembered one poem in particular from high school, "The Anecdote of the Jar." It was one of those perfect, embalmed poems that are most likely to be preached in school, and I had always held it in secret contempt. Now, however, forty feet underwater, I heard it again. It was no longer a finicky verbal still life but something authentic and urgent, a poem so precisely constructed that the logic of the universe seemed to sluice through it like the water of a diverted stream. "I placed a jar in Tennessee," it began,

> And round it was, upon a hill.
> It made the slovenly wilderness
> Surround that hill.
>
> The wilderness rose up to it,
> And sprawled around, no longer wild.
> The jar was round upon the ground
> And tall and of a port in air.

> It took dominion everywhere.
> The jar was gray and bare.
> It did not give of bird or bush,
> Like nothing else in Tennessee.

Though this part of the wall was called the Library, the promontory itself had no name. Like most features of the underwater landscape, it was anonymous. But this was a spot I knew I would return to again and again, so I borrowed a word from the title of one of Wallace Stevens's books and named it Harmonium Point.

I swam out to its farthest edge, hovered for a moment over the abyss and then, as if gravity were a threat, backpedaled with my fins until the solid coral rock was beneath me. Then I positioned myself vertically, set my hands on my hips, and let my heels touch down harmlessly on a patch of sand at the edge of the point. A small spotted moray eel rippled across the terrain in front of me, and near my foot was a sea anemone whose stout tentacles looked like a multitude of minature elephant trunks waving in unison. From the abyss, like a thought slowly forming, an eagle ray swam into sight. Its hide was speckled, and at first I saw only white spots, a dense, drifting constellation against the deep blue of the ocean. Then the rest of the creature slowly disengaged from the background.

Rays were common on the reef. I had already seen many stingrays wandering in the sand flats or skimming low over the coral surfaces. Stingrays are bottom feeders. They tend to snuggle into the sand and, when alarmed, take off in short, contour-hugging flights. But this was the first eagle ray I had seen, and an eagle ray is to a stingray as a Frisbee is to a cast-iron skillet.

Gliding and banking in front of Harmonium Point, the animal took on the power of an apparition. Its snout was elegantly tapered; its wingbeats were powerful and far more fluid than a bird's. Watching it, I felt my heart beat faster and felt a serene chill creep up my wetsuited spine. I thought that if I hovered here long enough, other such creatures would come to me: giant sunfish, sea turtles, basking sharks, humpback whales. My pressure gauge showed I was running low on air. A prudent diver surfaces with 600 pounds per square inch in his tank. I was a prudent diver, but I stayed on, sucking another few hundred psi and trying to sort out what I was feeling, what I wanted here. And it came to me that I simply wanted to be included, to be inside the frame for once. I wondered if that were possible, or if, as an alien visitor to the reef, I would always be like Wallace Stevens's jar in Tennessee, helplessly taking dominion everywhere.

4

Trees of Stone

What is coral? Diving on the reef among the towering boulders, the spreading disks, the waving fans and fluted branches, I thought from time to time I could detect something, a low background hum that told me the reef's awareness of itself went deeper than I could imagine. At particularly susceptible moments, when I was ranging deep along the wall and feeling a bit lightheaded from the effects of too much compressed nitrogen, it seemed to me that I could feel the breath of the sedentary living forms all around me.

A reef in its very topography is a living thing. The bedrock features of a coral reef — its boulders and canyons, its caverns and overhangs — are formed from the skeletons of coral polyps. The reef builds up layer by layer with each passing generation of these creatures, so that over time it becomes a mile-deep honeycomb of past lives. The shapes and colors we see covering its surface are only a fleeting veneer — a mat of polyps that will soon die and take their place in the inanimate substructure. While they are alive, though, they are among the earth's most beautiful and pestering creatures.

The mind does not easily embrace the idea that corals are animals. To the average person, coral is a variety of underwater plant, or some strange sort of flowering rock. Even Charles Darwin, who studied coral atolls while aboard the *Beagle,* seemed to pay scant notice to their biological dimension. He was fascinated with reefs as a geological phenomenon, and one of his earliest triumphs as a scientist was a book called *On the Structure and Distribution of Coral Reefs,* which answered the vexing scientific question of how coral atolls, those doughnut-shaped islands that dot the South Pacific, are formed. (Darwin demonstrated that atolls start out as fringing reefs on the shores of volcanic islands. As the island itself subsides, the reefs accumulate until finally the volcano has sunk from sight, and all that is left is a band of coral surrounding an empty piece of ocean.)

Darwin focused his attention on coral studies when the *Beagle* stopped on Keeling Island in the Indian Ocean near the end of its two-year voyage. By that time he was a weary young man, tired of chronic seasickness, lonesome for home, and perhaps worn down by his own revving curiosity. "We are all utterly homesick," he wrote to his sister, and a modern reader of *The Voyage of the Beagle* can still catch him in disagreeable moods. "It is excusable to grow enthusiastic," he writes dyspeptically, "over the infinite numbers of organic beings with which the sea of the tropics, so prodigal of life, teems; yet I must confess I think those naturalists who have described, in well-known words, the submarine grottoes decked with a thousand beauties, have indulged in a rather exuberant language."

Yet Darwin himself was enchanted by the mysterious

workings of coral, and when he began to describe them he could not strike the exuberance from his own voice.

> Let the hurricane tear up [the reef's] thousand huge fragments; yet what will that tell against the accumulated labour of myriads of architects at work night and day, month after month? Thus do we see the soft and gelatinous body of a polypus, through the agency of the vital laws, conquering the great mechanical power of the waves of an ocean which neither the art of man nor the inanimate works of nature could successfully resist.

Darwin knew those tiny "architects" were living creatures, but such knowledge had not been long in the world. For centuries coral was thought to be merely a festive inanimate feature of the underwater landscape — the "tree of stone." As certain curious people began to inspect coral more closely, to dredge chunks of it up from the bottom and watch it harden in the sun as the living tissues shrank and died, perceptions began to change. Coral was not a rock after all, but a flowering plant that bloomed nightly beneath the ocean surface. Naturalists began to refer to a coral polyp's secreted skeleton as "bark" and to the gelatinous polyp itself, when crushed between the fingers, as "sap." ("Very much like the milk from the breast of a woman," wrote one observer.)

Jean André Peysonnel, a French physician with a passion for natural history, put forth the notion that corals were animals (or at any rate "insects") in 1726. He sent a paper to that effect to the Paris Academy of Sciences, which received it with some mirth. The friend who presented the paper took the precaution of removing Peysonnel's name from it, afraid that the ridicule might cause irreparable damage to the author's reputation.

By and by, though, Peysonnel's unlikely assertion became

an indisputable, if still mysterious, fact. As I dove along the reef, as its odd creatures became as familiar to me as the squirrels and grackles of my Texas neighborhood, I continued to feel the vibrant strangeness of the coral. I could never really understand what exactly it was. Coral animals to my mind were a perplexing creation: somehow not quite living but not yet dead, not quite one being but not exactly many. They taxed the limits of my conscious understanding, and at the same time they hinted at unknowable but deeply felt notions, at the core sense we possess that our own lives are part of some common fabric. The lives of coral polyps, it seemed to me, were insensate, their tiny bodies a mere blob, but each individual lived and died in the service of some greater wisdom, in a blind striving for pattern, shape, structural mass.

"I was thinking the other day how beautiful the coral is," Mitch said as we filled the day's tanks in the compressor shed behind his house. "I was thinking, Okay — man's great civilization: he's got stop signs shaped like octagons, yield signs shaped like triangles. All these manmade shapes, but there's no pattern to the coral. So here's the concept: here's man's world of squares and angles and everything's messed up and hopeless. We spend so much effort trying to make things square or whatever, and nature just does what it does."

It cheered me a bit to think that after his thousand dives, Mitch still regarded the reef much as I did, as a chaotic, benevolent, overgrown garden. The coral blinded me with its variety, its seemingly random shapes and textures and colors, its immense patternless bulk. I had hoped to "learn" the reef and to have this expertise be a secret triumph, like the mastery of an obscure language one may never be called upon to speak. But I could not at first pick the pieces apart for study. I discovered I had the sort of mind that saw the life of the reef as a moving pageant, not an interlocking series of discernible tableaux. Nothing in particular seemed to happen before my eyes, just the intoxicating sweep of alien life: fish moving in morose glides or lightning lurches, gorgonians waving in the current, and the invisible, eternal respiration of sponges and tunicates.

Gradually, though, I began to see the reef as something comprehendible. The coral formations, which to my first bewildered surveys were as dense and indistinct as candle drippings, slowly sorted themselves out into recognizable zones. I was like a hiker in a mountain forest, dimly noticing how, as I rambled along, the landscape changed about me — the pine-scented lowlands giving way to dense growths of spruce and fir, then finally to a few contorted, wind-stripped trees at the edge of timberline. I began to perceive that on this point Mitch was wrong — there *was* a pattern to the coral. Like the trees of a forest, the various coral species were partitioned into climatic zones and spheres of influence. Near the surface, at the summit of the reef crest, stood the tangled branches of elkhorn coral. Farther below, but still flourishing in the light of the sun, grew massive coral boulders whose polyps were arranged in clusters or extensive winding channels. Deeper still, where the light was weaker, the boulders flattened into fanlike forms, their individual polyps spread out like the cells on solar collectors to better absorb the diminished sunshine. And those were only the most noticeable varieties. There were scores of others, hard as rocks or soft as ferns, waving listlessly in the current or standing as bulwarks against the powerful surge. Some of the coral polyps were nearly the size of saucers, others were mere specks, pitting the surface of formations that, it seemed, could take just about any form: a palsied, grasping hand, a candelabra, a spreading fan, a bright yellow loop of electrical wire.

A coral polyp is a writhing dab of protoplasm sheltered within a rigid cup. At one end it is equipped with a ring of tentacles, which the polyp extends beyond the lip of the cup to seize prey. In the center of the ring is a mouth, an all-purpose seam through which the polyp receives food, expels waste, and — in the spectacular and mysterious annual eruptions of certain species, timed precisely to the phases of the moon and the workings of the tides — releases vaporous clouds of sperm, and eggs in the form of bright orange globes. I've never seen this secret moment — few people have — but in photographs it's unsettlingly lurid: the soft mouths of the polyps puckering to release the perfect spheres of their eggs, the funky background mist of floating sperm, a general air of ardor and abandon that is at once ungraspable and familiar.

Those photographs came back to me as I swam along the reef, watching the butterflyfish pluck food out of the coral with their thin snouts, watching mating pairs of blue tang spiral off from their schools, heading toward the surface in a nuptial fury so intense that in an instant the male had turned white, his royal blue skin blanched with what I guessed was desire. I was haunted by one thought: what do other things feel? Could there be a dimmer bulb of awareness than a coral polyp? And yet how could it not be bristling with a sense of its own existence, how could it not quiver with satisfaction — a satisfaction primitive to us but profound to itself — as it felt the release of those eggs out of its body and into the water? Looking at those photographs, I thought that what a coral polyp experiences at that moment might not be, at its core, so distinct from a human orgasm — a shuddering triumph of destiny fulfilled, a satisfaction deeper than thought at having put into circulation the material for a new generation.

Sometimes I felt like a lordly presence on the reef. Sometimes I felt *lucky* not to be a thoughtless, flailing thing like a coral polyp or a creature as indifferent as a goatfish, joylessly probing the sand with its fleshy whiskers. But more often I was swayed by the opposite conviction, proud of my insignificance, happy to be just another component swallowed up in the oneness of being. I could feel that old human arrogance falling like a weight. All my life I had believed, without realizing it, that the interior lives of other creatures mirrored my response to them. If they were repulsive to me, they lived in a state of self-loathing. If I did not find them particularly interesting, they inhabited a kind of limbo in which their stunted souls would never find full expression. The idea — the dawning conviction — that these slimy, blobby, indistinct life forms were no less aware or alive than a human being, that there is finally no scale on which the relative value of creation is measured and recorded, filled me with a kind of spiritual consolation.

Spotty clouds of plankton continually drifted through the clear water and settled on my wetsuit like dandruff. I could see the white grains of living matter surging about in the waving hair of my forearms. Like so much else in the ocean, a coral polyp begins its life this way. Launched through the mouth of its parent, the larva is a waterborne speck, fringed with hundreds of invisible beating hairs that move it along with the rippling motion of a caterpillar. In consort with billions of other planktonic larvae it drifts through the blankness of the sea like an unmanned satellite, silently heeding its onboard instructions. This unformed polyp — it's called a planula — may by some fantastic chance avoid being lethally strained through the body cavities of comb jellies, shrimps, fanworms, sponges, crabs, or mature corals. It may escape the currents that would sweep it out to sea and send it drifting hopelessly down into the cold ocean depths. Instead it sinks down upon the relatively shallow, sunlit surface of the reef. It settles on a fragment of rock or on a ship's anchor or on the abandoned engine block of a 1954 Oldsmobile Cutlass — it knows what to look for, and will die if it does not settle on suitable terrain. But if the planula manages to locate a congenial substrate, almost at once it begins to seethe and change shape, growing into a stubby tube with its ring of tentacles and simple slit of a mouth.

It walls itself in, extruding calcium carbonate to form an exterior skeleton whose shape — as precise in its planes and angles as a military redoubt — is decided by the genetic blueprint of its particular species. The sheltered creature then pulls itself apart, budding off another individual, which begets another, until in time that single pioneer polyp has transformed itself into an entire colony. Each polyp is a separate being, living in its own cup, but it is connected to other members of the colony through a blanket of mucusy tissue. Onto this blanket drift various forms of protozoa and nanoplankton — prey too small to be grasped by the polyps' tentacles. But they stick to this surface and are delivered to the mouths of the polyps on a swift tide of mucus.

In my mind a coral colony was like an apartment building made up of identical units, each one holding a solitary tenant. (I imagined these tenants as bored and listless, joylessly studying the labels on soup cans.) But this is a defective image, because the tenants are not solitary, they are *connected*, bound to each other by shared tissue. They are in some sense one continuous creature.

Corals are not just passive harvesters of the reef's bounty. They are predators as well. Coral polyps feed on plankton, organic debris, worms, and even small fish. On most of the big reef-building corals, the polyps spend the days secreted in their cups, but the lure of moonlight or the chemical taste of prey in the water pulls them out. Imagine that apartment building. At a given moment at the end of the day each of its many tenants grows suddenly alert and raises his head to sniff the air, shaken out of his stupor. Each walks out onto his balcony. A blizzard of food is drifting through the air — doughnuts and rump roasts and corn on the cob — and the tenants, in unison, their feet planted firmly on the balconies, lunge for it with extended arms.

In grasping for food that is alive and wiggling, coral polyps have the advantage of six or more arms. The tentacles are covered by microscopic threads that function like the trip wires of booby traps. When a passing creature brushes these hair-trigger threads, venomous coils shoot out. The coils of one species ensnare the prey like a whip. Another kind attaches itself with a deadly adhesive. Still another, the most common, embeds its barbed end directly into the victim's body. Trapped and subdued, the creature is then delivered to the polyp's gaping mouth.

As predators, corals are voracious and efficient. (They are also protective of their hunting grounds. If rival colonies begin to intrude on their patch of reef, the polyps attack with chemical barrages, with specialized tentacles, or with lethal digestive filaments that slurp away the invader's tissue.) Most of the hard, reef-building corals hunt only at night. At daybreak the tentacles stop writhing and the polyps shrink back into their shelters. But the coral's food factory is still on line; there is merely a changeover in equipment. Each polyp harbors an algae plantation, many thousands of single-celled plants called zooxanthellae. In magnified photographs, the zooxanthellae are visible as dark, malignant-looking smudges floating within the polyp's clear flesh. They are not parasites, however, but providers. The algae thrive on the polyp's waste products — nitrogen-rich ammonia and carbon dioxide. In return, the animal host receives oxygen and nutrients that the algae have produced through photosynthesis.

At any moment on the daylight reef, this is the great unseen process: sunlight raining down on microscopic zooxanthellae, poison gases changing into beneficent gases, simple sugar molecules appearing like dew on the insides of coral polyps. Without those little algae, the reef-building corals would die.

One day Mitch and I were swimming through a broad rift in the wall at about sixty feet when we came across a group of divers. Mitch introduced me to one of them, a marine biologist, by writing her name on my slate: Cindy. We shook hands.

Cindy borrowed my slate and wrote on it with the stubby carpenter's pencil attached to it with nylon cord.

"Have you noticed," she wrote, "the bleach white of the plate corals on wall due to loss of symbiotic algae?"

"No," I responded, in my inferior underwater penmanship, "but I happen to have a questionnaire on that subject I'd like you to fill out."

She looked puzzled when she read this, and I could think of no way to explain to her underwater that a marine biologist from Puerto Rico, having heard that I was planning to spend some time in Grand Turk, had mailed me a stack of questionnaires on the bleaching problem to hand out to everyone I met. All over the Caribbean that year, corals had been mysteriously casting the zooxanthellae out of their bodies. In some places the process was so intense that clouds of expelled algae turned the water brown and turbid. Zooxanthellae not only manufacture the bulk of a coral polyp's diet, they provide it with color as well. So as the plants floated off into the water, they left behind whole tracts of reef in which the corals were as white as ivory — and slowly starving. No one understood exactly why this was happening, though it appeared to be related to a rise in water temperature and was interpreted by some scientists as a harbinger of the effects of global warming.

I followed Cindy as she led me on an inspection tour of the bleached coral. The reef at this location was laced with canyons and tunnels, and we followed them — dropping down to eighty feet or so — until we emerged through a hole in the face of the wall. Here she pointed out five or six large plate corals, growing out from the wall in flat arcs like giant tree fungi. Most of them were bone white, and a few had a sickly brownish cast.

Cindy waved goodbye and left to join her group. I never saw her again. Though I was growing low on air, I spent a moment or two more among the ailing coral formations, pressing the lens of my mask close to their surfaces so I could see the colorless polyps shuttered up inside their silos. If the water cooled a bit, they would regenerate their store of zooxanthellae; if not, they and the other reef-building corals might die.

And bleaching was only one of a host of problems that would afflict them in years to come. Smothering loads of sediment from coastal construction and inland erosion; pollution in the form of pesticides, heavy metals, and bacteria; inexplicable population surges of ravenous, coral-eating starfish — all of these catastrophic events were looming or had begun to arrive in force. I felt a surge of pessimism, fueled in part by the inability of my imagination to hold these creatures in focus. Like those smug naturalists in Peysonnel's time, I was not sure that I truly believed them to be animals. *Animal* was too bold a word for such ambiguous and perishable life forms; they seemed to me as fragile as flowers, capable of withering in my sight.

5

Tierra

In the open-air bar of the Salt Raker Inn a heavyset folk singer from Liverpool pounded the strings of his guitar with a closed fist. He had a ferocious voice, and he sang the chorus as if he were declaiming some urgent self-revelation:

> Fer I'm a ROVER!
> Fer I'm a ROVER!

At a table near the back of the room, next to a wooden pillar festooned with fish identification charts, I sat with the two Bobs.

"You know why they're called the chosen people?" said the older of the Bobs. He was in his seventies, thin as a wolf, with mean, narrow eyes and a small white handlebar mustache that he twirled like a stage villain to enhance his outrageousness. "It's because they were chosen by Satan. Look it up — John eight, forty-four. Jesus says, 'Ye sons of Satan.'

"Incidentally, I can prove that Jesus was not a Jew. It's a

simple concept to grasp. As you know, Mary being ever-virgin, Jesus had to have been born by artificial insemination. Mary was his surrogate mother — nowhere in the Bible does it say that Jesus was born from her egg. It was God who put the fertilized egg — *fertilized* — into her womb. And was God a Jew? Hell, no!"

The old man cackled in triumph, took off his gold-braided captain's hat, and patted down the half dozen strands of hair arching across his bald head. The younger Bob gave a patient, sardonic smile, awaiting his own moment to shine in conversation. He was fifty and rather urbane, with a mustache as discreet as his friend's was intrusive. On his finger he wore a heavy gold ring in the shape of a lion's head. The two Bobs had come to Grand Turk, I gathered, with some vague idea of setting up an offshore trust business. They had also dropped a few hints that they were trying to track down someone who had cheated them on a business deal and who was now hiding out from them in the islands.

That morning they had gone out on Mitch's boat to snorkel, and I had spent part of my dive watching them from below as they swam about on the surface. For all their hateful haranguing, they had an awkwardness in the water that made them seem fretful and helpless. Bob the Elder stroked about with his spindly limbs like a waterbug, following the progress of the scuba divers as we filtered through the terrain below him. His companion stayed near the boat, frantically paddling and gulping water through his snorkel.

"Did you see the jewfish?" Bob the Elder announced as we climbed back onto the boat. "I don't know why the Jews let us get away with calling them that. Hell, they're even trying to change the names of all the cities named after the saints. San Francisco, Saint Paul, San Antonio — they won't rest till they've changed them all."

That was the only comment either of them had made about their underwater experience. Bob the Younger, especially, seemed not to have cared for it. Perhaps the fact that he had no ready opinions about the ocean — no way to quell its threat with his ruthless logic — had left him unsettled.

In the bar, however, he was happy to rattle on. As a matter of principle, he told me, he didn't pay taxes or even obey laws. He manufactured his own license plates, which he inserted into a custom-made dealer's frame that said "Free-man Motors."

"I have rescinded my social insecurities," he announced. *Sprechen sie Deutsche?* No? Well, there's a wonderful German saying" — he quoted it fluently — "which translates to 'I have had enough.' Let me put it this way. When you're born you're issued a birth certificate, right? Where is a copy of that birth certificate sent? The census department? No! The commerce department! From the day of your birth you are assumed to be engaging in international commerce! It is up to you to say, 'I object.'

"For instance, when a policeman stops me for some petty traffic violation, I remind him of the three elements that must be present before a crime can be said to have been committed. One, actus rei — the act itself. Two, mens rea — the presence of evil intent. And three, corpus delicti. I get out of that car and look under the tires and say, 'Officer, do you see a corpus delicti here?' "

I shifted position in my chair. I had been underwater for most of the week, and the muscles in my upper back ached from the weight of the scuba tank. My mouth was sore from gripping the mouthpiece of my regulator, the skin of my toes had been rubbed raw from chafing against the hard black rubber of my fins, and I was aware of a persistent, nagging sting on my arms where I had brushed against a growth of fire coral. And all the bilge and bigotry from the two Bobs registered as a kind of physical discomfort.

"Before you go," Bob the Elder said, as I stood up to escape, "have you ever seen a dollar?"

"A dollar?"

He took an 1881 silver dollar out of his pocket and dropped it on the table.

"Now that has a ring to it," he said. "What people don't realize is that a dollar is a unit of measure — like an inch or an acre. That dollar there is four hundred and twelve and a half grams of standard silver, troy weight.

"Now what's this?" He set an ordinary dollar bill on the table. "Does this look like four hundred and twelve and a half grams of standard silver, troy weight? Does it? Well, does it?"

"No."

"So what you're saying is, one of these is a phony. Is it the silver dollar, or is it the piece of paper? Come on, tell me which is the phony?"

"The dollar bill," I said wearily.

"That's right!" he announced triumphantly. "And do you know who we have to thank for it? Franklin Deficit Roosevelt!"

I walked down Front Street, along the edge of the sea. The night was clear and still, the ocean a heaving shadow interrupted only by streaks of phosphorescence and the running lights of a few boats. It was Saturday night, but downtown Cockburn Town was almost deserted. A light westerly wind coasted in off the water, driving six-inch waves that slapped softly against the pilings of the piers. I could smell sea wrack, garbage, decay from the salinas, and deep-fat frying from the Poop Deck, the restaurant down the street. In the center of the street a wild donkey brayed urgently, a sound like a creaking hinge swung violently back and forth.

From the open doors of St. Mary the Virgin Anglican Episcopal Church, a forthright structure whose white limestone facade almost gleamed in the darkness, I could hear the sounds of choir practice, a host of black voices singing the sonorous hymns of faith of the British Empire. Listening to the hymns and the braying of the donkey, I walked out onto a rickety pier and stood there gazing seaward or, rather, down into the dark water. Even after weeks of constant diving, the world beneath the surface still seemed utterly secret, my memories of it constantly threatening to vanish. The reef itself, not just the water that covered it, was a fluid presence in my mind. When I was on the surface, Harmonium Point presented itself to my imagination like a taunting memory of childhood, like some moody, mildly exotic place glimpsed through sleepy eyes at the age of three or four — a place that you will never see again or even clearly remember, but that will always call mysteriously from some deep chamber of your mind.

It always surprised me a little to find Harmonium Point and the other landmarks of the reef still there when I slipped over the side of the boat. To see these places again was like going to sleep and reentering the landscape of a dream from the night before. Out of the water, out of that dream, I was habitually restless. Tonight, as always, I feared I was missing something. I was anxious about everything that was going on without my knowledge beneath that dark water.

I walked through Cockburn Town, listening to the Anglican choir sing "My Father's God to Thee," and then climbed onto my scooter and rode around aimlessly in the night. There was a thin slice of moon in the sky, with two wispy tendrils of cloud around it like parentheses. A solitary horse clopped down the road. I passed the prison with its rolls of concertina wire crowning a limestone wall, and farther down the road a neighborhood of ramshackle houses and bars, some listing at pronounced angles, light and music pouring out from the gaps between the boards.

I ended up on the opposite side of the island, standing on the summit of the limestone ridge looking out at the barrier reef to windward. I could see the faint line of surf beyond the lagoon, and the scant moonlight softened the raw gray rock on which I stood.

Christopher Columbus, sailing through the Caribbean at two A.M. on October 12, 1492, bearing among his other effects a letter of introduction to the Grand Khan of China, first saw the Indies as a pale moonlit cliff such as this. I had spent the afternoon with a man who claimed that it *was* this one, that Columbus's first landfall had not taken place, as generally supposed, at Watlings Island or Samana Cay in the Bahamas, but here at Grand Turk.

"He would have been coming right across from where we're sitting now," Herbert Sadler told me as we drank tea on his porch overlooking the windward side of the island. Brown pelicans flew teetering past us at eye level, looking as primitive as pterodactyls. "He would have seen the white limestone cliffs, he would have seen the waves breaking over the reefs. The night he came here there was a strong northerly wind blowing, so his ships would have been blown to the south of Grand Turk and anchored in Reef Harbor, which is the most perfect anchorage in the Bahamas."

Herbert Sadler was the island's historian as well as its most prosperous grocer. I had met him a few days after I arrived, when I stopped into his grocery store to buy his four-volume history of Grand Turk.

An old man with no apparent hair, Sadler was sitting on a

chair in the middle of the sales floor as a barber snapped a pair of shears in the air around his head.

He was gruff but polite, eager to educate me about the theory that Grand Turk was Guanahini, the island where Columbus made his first landfall. I listened all afternoon to his evidence — how Grand Turk fit exactly the description of the island in Columbus's log; how a fifteenth-century nao or caravel, sailing before the westerly trade winds from the Canary Islands, would almost certainly slip into the southerly currents leading toward the Turks and Caicos; how Grand Turk was the only island in the vicinity of the northern Bahamas from which Columbus could have seen, as he noted in his log, "so many islands that I could not decide to which I would go first."

"We think we have answered every possible angle," Sadler said. "We think that if the final judgment is made by a reasonable authority, we will win hands down."

I had listened without skepticism. The site of Columbus's first landfall has never been conclusively fixed, and a score of islands in the Caribbean, each with obsessive local historians armed with nautical charts and tree-ring dates and weather reports from 1492, claim the distinction. With so much contention and confusion on the part of scholars, it seemed like a harmless indulgence to go along with Sadler and believe that this was the place — that the meager bluff on which I stood was what the lookout on the *Pinta* had seen when he cried out, no doubt at the top of his lungs, "Tierra! Tierra!"

Today Grand Turk is a desert, a looming shelf of rock covered with chaparral and skittering lizards. When Columbus saw it — *if* he saw it — it would have been crowned with trees, and the parched salinas in the center of the island would have been a gleaming lake. And it would have been populated by a people now extinct, thanks to the almost unimaginable rapacity Columbus willingly loosed upon the New World.

They were the Island Arawaks, called by the Spanish the Tainos. They greeted Columbus with something like joy, and he in turn appeared beguiled by them. "I assure Your Highnesses," he wrote to the king and queen of Spain, "that in all the world there is no better people. . . . They love their neighbors as themselves, and they have the sweetest talk in the world, and are gentle and always laughing."

The Tainos wore no clothes, which struck the explorers as charmingly licentious. They lived in roomy conical houses and slept on hammocks, and planted manioc and sweet potatoes in built-up mounds of soil. When Columbus and his men sailed on to Cuba, they found the Tainos there indulging in a curious custom — walking about "with a firebrand in the hand and herbs to drink the smoke thereof." The Indians not only smoked tobacco — something European eyes had not yet witnessed — but they chewed it as well, possibly mixing it with a narcotic herb that induced hallucinations.

One drawback of Herbert Sadler's theory is a lack of archeological evidence to support the idea that Tainos were living on Grand Turk when and if Columbus arrived. As far as is known, the Indians did not have permanent settlements on the island, but visited there sporadically to collect shells for the manufacture of beads. There were hundreds of thousands of Tainos, probably even millions, living on Hispaniola and Cuba and the smaller islands like Grand Turk, ranging throughout the Bahamas in their canoes made of carved and burned-out tree trunks. By 1542, fifty years after Columbus first sailed into the Caribbean, there were hardly more than two hundred Tainos still alive.

"They should be good servants and intelligent," wrote Columbus, in one of the most chilling remarks in American history, "for I observed they quickly took in what was said to them. . . . They are good to be ordered about, to work and sow, and do all that may be necessary."

And that is what happened to them. They were kidnaped to dig gold mines and to work as pearl divers in Venezuela. They were borne away by smallpox, sold into slavery, torn apart by Spanish hunting dogs, burned alive, hanged, and used by soldiers for swordsmanship practice. "A ship going to Hispaniola to the islands," wrote Bartolomé de Las Casas, the priest who became the howling conscience of the Spanish conquest of the New World, "sailed whither without any compass, only by the bodies that floated up and down the sea."

The last Taino to see Grand Turk was possibly the one encountered by Ponce de León in 1512, when he stopped at the island to enlist guides in his search for the island of Bimini and its Fountain of Youth. Since there was only one old man there to help him, Ponce de León gave the island the name Del Viejo.

For many decades afterward Grand Turk remained unpeopled, though not unvisited. In the late sixteenth century English settlers, sailing across the Atlantic to establish their colonies in Virginia, stopped at the Turks and Caicos to take on salt, to hunt manatees (now long gone from these waters, as from most everywhere else), and to exterminate great numbers of flamingos, which they called "flaming swanees." The island became a haven for French buccaneers and for the far-ranging Moroccan pirates known as the Salee Rovers. In the 1670s, enterprising colonists from Bermuda began voyaging south to the Turks and Caicos with their black slaves to rake salt from the broiling salinas.

Grand Turk was fertile in salt. The shallow ponds in the

middle of the island, baked by the sun and swept by offshore winds, were a natural factory, the briny waters melting into the air to expose beds of white crystal. Those first Bermudian entrepreneurs lived in makeshift huts and subsisted on salt pork and iguana meat. They chopped down trees to reduce the shade and hasten the evaporation of the water in the salinas, and when the salt was ready they raked it with long wooden rakes into white mounds. (These mounds were later commemorated in the official Turks and Caicos flag, which flew over the colony for a hundred years until its iconography was redesigned in 1967. The flag featured one glaring mistake: the designer in distant London who drew the final sketches added doors to the salt mounds, under the impression that they were igloos.)

Turks Islands salt was remarkably pure, and the Bermudians found a ready market for their product among New England fishermen, who needed salt to preserve the cod they exported to feed the slaves laboring in the Caribbean's sugar islands. During the American Revolution, many Bermudian merchants — traitors to the Crown — risked the British naval blockade of the colonies to supply salt to Washington's armies. But after the revolution the Turks and Caicos, along with the rest of the Bahamas, were soon overrun by outcast American Loyalists.

The cemetery at Grand Turk is filled with the broken crypts of those salt rakers and dispossessed Tories, with faded stone tablets "sacred to the memory" of the young wife of a colonial secretary or the baby daughter of an officer in the West Indies Regiment who succumbed on this distant, sun-blasted speck of the British Empire. There is a hint of despair in the old cemetery — everything bleached and broken, the lettering on the tombstones eroded, lizards and land crabs crawling in the shade of the open vaults. And the inscriptions seem despairing as well, speaking not of hopeful eternal life but of "affliction," "bereavement," "untimely and irreparable loss." Wandering through the cemetery, I sensed that when these people were laid in their graves Grand Turk was already far, far removed from the jungle paradise of the extinct Tainos. It was a lonely, parched wilderness, an island that had been scraped clean of life and then sown with salt.

The names on those tombstones — Lightbourne, Misick, Williams — still survive in Grand Turk today, but the people who bear them are more likely to be the descendants of African slaves than of British pioneers. The Bermudians brought slaves to Grand Turk; so did the Loyalists who followed them. Until the British outlawed slavery in the early nineteenth century, entrepreneurs from the island traveled to the markets in Nassau and Santo Domingo to buy slaves to work in the salinas and in the short-lived cotton industry, which was ruined by boll weevils. Other Africans ended up in Grand Turk when slave ships were wrecked on the reefs or liberated by the British navy. These unfortunate people — captive West African tribesmen imprinted with the horrors of the Middle Passage, condemned to a lifetime of soulless servitude — gave rise to the majority population of today's Grand Turk, black men and women whose accent, in its tendency to substitute "w" for "v," still bears traces of antique English.

Herbert Sadler's authoritative history of the islands is a home-grown affair, illustrated with badly reproduced maps and photographs and packed with dense, faded type. I could not read it for long before succumbing to eyestrain and a certain intellectual listlessness. By and large the history of Grand Turk is a litany of things that never quite came off: a battle between Lord Nelson and the French that was never joined, a cotton industry ruined by boll weevils, revolutions that were averted, pirate treasures that were never found.

Under the water, in the reefs surrounding the island, there was no history at all, unless you could count the ceaseless spawning and dying of ocean creatures and the steady accumulation of sand and coral rock. I looked down into the sea from the bluff, watching the waves collide against the rock, hitting its hollow pockets with a sharp clap and sending up a plume of spray. It was late at night, but the desire to be underwater was irresistible. I jumped on the scooter and rode back to my hotel room at the Island Reef, gathered my masks and fins and an underwater light, and walked down to the beach.

6

Eye to Eye

Hummocks of dried turtle grass, uprooted from the lagoon and washed ashore by the mild play of the waves, stretched all along the beachfront. Walking through the grass in my neoprene booties, I could feel shells and crustacean parts crunching beneath my feet. The water was warm enough — no need for a wetsuit if I kept moving — but the lagoon was shallow, and I had to walk for twenty or thirty yards in knee-deep water until the terrain began to slope enough for me to submerge. I slipped my face beneath the water and looked around before I turned on my light.

"The book of moonlight is not written yet," wrote Wallace Stevens, "Nor half begun." I was in a landscape of soft steely light, the moonlight washing through the waves and illuminating the sand flats and the distant coral boulders with sumptuous hues of gray. I could hear the wind whistling through my snorkel. I could hear the distant braying of a donkey. I could feel my body rocking gently, buoyant in the waves. The world was colorless but welcoming, an enfolding darkness full of unexpected texture and richness.

After a moment I turned on the light. The beam fell on a waving bed of turtle grass, from which thick, fan-shaped growths of algae protruded. Here in the grass beds and sand flats there was little color, even with the light beam to coax it out. The grass and algae had merely a dark greenish tint, the sand was white, the occasional sea cucumbers as brown and russety as baked potatoes.

Swimming out into deeper water, I took a breath and dove to the bottom, cruising above the ripples in the sand or following snail tracks to see where they might lead. Tiny, frantic fish skirted around me, and when they crossed the beam of the light their shadows were projected onto the sand at many times their normal size; I felt as if I were surrounded by some undefinable fluttering movement, like the anxious beating of bat wings.

I had dived often at night, though always with scuba gear and never alone. But snorkeling now through the dark sand flats made me feel exposed and edgy, and whenever I came up for air I was struck with how much more vulnerable I felt here at the surface than below. With a scuba tank on my back I always had the sense of being armored and full of options. I could hide behind a rock, wedge myself into a crevice until whatever it was that had scared me lost interest and swam away.

And what *was* I scared of tonight? Not sharks, I bravely told myself. The days are past when recreational divers, upon spotting a shark in the water, would scramble up to the boat in a froth of panicky bubbles. Nowadays divers swim toward the sharks, wanting a better look, proud of their educated fearlessness. We are attracted to sharks because they are one of the few mythic dangers of the sea that have not been thoroughly discredited. Experienced divers know that octopuses are not ferocious, that giant clams do not routinely close down upon human feet and hands, that moray eels do not lunge from hidden crevices to grab you by the neck and slice open your jugular vein with their teeth. But about sharks we are never sure, so we constantly crave the sight of them, wanting a little harmless jolt of terror.

Or perhaps not so harmless. Sharks do attack humans, of course, though as predators they are far more specific in their targets, far more clear-headed, than we usually give them credit for. For most of us, a shark is appetite incarnate, a thoughtless greedy maw roaming through the ocean, an ingesting machine. Its seeming indifference — reflected in its flat eyes, the scowling set of its mouth, the mechanical sweep of its tail — frightens us, because that indifference is a measure of the utter pitilessness with which we imagine the shark regards its victims. In our eyes, sharks have no intelligence, just a brutal will.

In fact, they are unusually perceptive fish. Their brains are large, and in laboratory experiments they have shown an ability to respond to simple commands and even to retrieve objects thrown into their pools. Although most sharks, like most fish, are cold-blooded, a few species — among them the great white — have been found to have warm blood. (To determine this, an inventive shark researcher took a knife and made an incision in a captive great white and then stuck a thermometer into the wound.) Sharks have good eyesight, and through unusual receptive organs on their snouts, known as the ampullae of Lorenzini, they are able to detect exquisitely faint electrical pulses. A researcher named A. J. Kalmijn noted that a shark would be able to sense an electric field "equivalent to the field of a flashlight battery connected to electrodes spaced 1,000 miles apart in the ocean."

Most shark attacks on humans seem to be cases of mistaken identity or misread signals. Seen from below, a surfer in a wetsuit, his limbs hanging over his board, offers a convincing likeness of a seal. And a big proportion of shark attacks are on spear fishermen, as they drag their thrashing, bleeding trophies to the surface. Homing in on the signals given by the fish, an attacking shark may instead grab a neoprene-clad leg or arm.

I was trailing no fish tonight, but my imagination was at work and I was growing wary. I remembered the urgency I had felt long ago on my first few dives in the open ocean, surfacing and then pulling myself up onto the diving platform with a slightly panicky vault, making sure I did not trail my legs in the water, where a beast could rocket up through the dark ocean and grab them. I no longer worried about sharks that way, but I was thinking about them tonight, scouting the water with my eyes. I imagined a gaping mouth, blacker than the ocean blackness surrounding it, as black as a collapsed star, sweeping in with no sound or order to announce it, just a sudden concussive turbulence in the water.

Shark attack victims rarely see the creatures that bite them. There is just that stunning blow out of nowhere — like being hit by a train, one survivor described it. I thought about the thirteen-year-old Australian boy I had read about, seized by a shark while swimming in New South Wales. Six rescuers heroically entered the bloody water to save him, thinking the shark was gone. But as they hauled the boy to shore, he kept shouting, "Get it off! Get it off!" Beneath the waterline, hidden by the murky water, an eight-foot shark was still attached to the boy's leg and would not let go. They carried the shark along with its victim to the beach, and pounded on its head with a surfboard until it released the leg.

Of all the shark attacks I had read about, that was the one that chilled me the most. The boy had recovered, but I wondered how he had dealt with the memory of that sudden horror, the innocent ocean congealing into a pair of jaws and seizing him in a relentless and specific grip, as if no other victim would do.

It was time to tone down my imagination. I had fallen into a suggestible mood, and I thought I could feel things cruising about in the water with me, just out of the range of my available senses. It was as if the ocean itself were a supremely sensitive trap and every move I made set off a vibration that alerted the unseen beast residing at its center. There was no danger, just the uncomfortable closeness of the night pressing in.

Many of the daytime herbivores were asleep now, or as close to sleep as fish get: holed up in a crevice, canted motionlessly to one side, their minds locked in some monotonous waking dream. But the carnivorous night shift was out — the cardinalfish and squirrelfish with their big, liquidy, night-perfect eyes; the spotted morays rippling through the coral underbrush; the lobsters prowling along the bottom with ghostly, delicate movements of their many limbs. I saw a deep linear gouge in the sand, like the tread of an off-road tire, and followed it for a while, hoping to find the sizable stingray that had made it. But I lost the trail when my attention was diverted by a porcupinefish. These boxy fish, covered with spines, have the sort of mild, expressive features you see on the faces of the valiant trucks and tugboats in children's picture books. Porcupinefish, along with puffers, which resemble them, are inflatable. They can instantly gulp enough water to expand like a balloon, driving their spines into the mouth of an attacker.

Had I been feeling bothersome, I would have provoked this one into inflating itself, but I was careful about my manners tonight, nervous about the dark sea brooding in judgment over my every move. And I liked watching the porcupinefish swim about in its unexcited state, its fins whirring furiously to keep its body upright. It moved seemingly without volition, teetering and tilting as if it were powered by a gyroscope.

As an obvious food source, puffers and porcupinefish have all the appeal of a backyard toad. Their flesh also contains a poison, tetrodotoxin, that is more often deadly than not. A careless diner can expect dizziness, nausea, numbness in the hands and feet, followed by a creeping paralysis that finally shuts down respiration. The Japanese, of course, are famous for eating puffers, known as *fugu*. They are prepared by licensed chefs who expertly (one trusts) cut away poison-free slivers of flesh and arrange them into glistening edible mosaics depicting herons and ornamental fish. The Japanese also make wine from puffer testes and unthinkable snacks from the ovaries, which require years of pickling before they're safely detoxified.

As I snorkeled closer to a scattered series of coral heads, the smooth sand bottom gave way to rubbly terrain, and I saw an octopus moving like a stream of mercury across the rocks. The octopus was a small one — a foot and a half, probably, from arm tip to arm tip — but it was such a restless, slithering mass that thinking of it in terms of size seemed irrelevant. Like a gas, it could be measured only by estimating its volume.

I took a breath and dove down to it, watching it change

colors as it slid over the rocks. In an instant it had turned from white to lavender to a warty institutional green. Its head was floppy and bulbous, like some grotesque tumor, and its eight radiating arms were joined for half their length by a rippling membrane of flesh. The octopus had not seen me and was not in a hurry. It proceeded with elegant fluidity, flinging one arm after another forward in a movement that reminded me of the slow-motion cracking of a bullwhip. Unlike their squid cousins, octopuses are not particularly adept swimmers; they're crawlers. They move by probing with their arms and gaining purchase on the substrate with their thousands of powerful suction disks.

They are capable, however, of sharp bursts of velocity. When I reached down and touched this one, it instantly turned brick red, let out a cloud of ink, emitted a blast of water from its funnel and shot upward, its eight legs trailing like streamers from its wobbling head. It landed a few yards away and poured itself into a hole no more than a few inches in diameter. One arm then flashed out of the hole, twirled around a nearby beer can, and drew it in front of the opening to block my approach. I pulled on the beer can to test the octopus's grip. It felt as if it were cemented into place.

The beer can did not totally cover the entrance to the creature's lair, and when I shone my light through the tiny opening I could see, glaring out at me from the darkness, one of the most awesome sights in all of nature: the eye of an octopus.

An octopus is an invertebrate. It is a mollusk, like a garden slug or an oyster. But unlike most other mollusks, which "see" only through blotches of light-sensitive pigment, an octopus has a startlingly complex eye, complete with iris, retina, pupil, lens, and cornea. The octopus eye, we assume, sees more or less what the human eye would see, and like the human eye it not only receives but emits. It takes in light, forms images, and sends out expressions of mood and thought. When you look eye to eye at an octopus, something clicks. You feel it staring at you, taking your measure, *thinking* about you. There is a disturbing, almost unspeakable recognition. You realize that the octopus's eye, like yours, is the window to its soul.

To its soul? Perhaps this eye, with its weird familiarity, its brooding awareness, merely deceives us into believing that the octopus has a reverberant consciousness to go with it. But it is no illusion, I think, that the octopus has a temperament — and that it is governed by fitful storms of emotion so powerful and sweeping that they seem to emanate from some larger source than the compressed knot of tissue that serves as the creature's brain. An octopus is easy to rile, easy to scare. It pines, it rages, it harbors grudges, it plots its moves. It announces its state of mind by instantaneous changes of color, the muscle-activated pigment cells in its skin compressing and expanding like the pixels on an electronic scoreboard.

I remembered an octopus I had encountered several years earlier, when I was diving off a boat in the Gulf of Mexico. I had been sitting on deck, waiting to go down again, when one of the divers surfaced, cradling an octopus in his arms like a baby. When he got into the boat he filled a deep ice chest with sea water, put the octopus inside, and jumped overboard to resume his dive. I was the only person on deck. The diver had said nothing about his plans for the octopus, and since I was fairly sure they were not in the animal's best interests, I thought about putting it back into the water. I was still thinking about it when I saw a tentacle drape itself over the edge of the ice chest. Very slowly and stealthily the rest of the octopus followed. Balanced on the Styrofoam rim, it paused and moved its eyes. I saw it looking at me, not a blind stare but a knowing assessment of risk. The octopus was taking stock of its situation. Finally, it poured itself onto the deck and, without hesitation but without evident hurry, slithered to the transom of the boat and calmly disappeared.

This octopus, the one I had cornered now in its lair, regarded me with the same air of calculation. It was thinking: *What will happen to me now?* But it had calmed down, subsiding to a mottled gray hue, more or less secure in its home base. Octopuses are extremely attached to their dwellings. "The octopus's ventral arms," writes Jacques Cousteau, "seem to be used more for anchorage than for any other purpose. One of the main preoccupations of the octopus is to maintain a point of contact with its rock or its hole. From this contact, it derives strength and confidence."

Octopus holes are clean and smooth inside, swept clean of debris by periodic blasts of water from the octopus's funnel. Outside they are junkyards, littered with old crab parts, bits of coral, or other objects the octopus has collected or cast off. Sometimes the octopus walls up the opening, building a rampart of shell or stone. Its need for a close, smooth, secret place into which it can pour its liquid body is intense. Octopuses routinely take up residence in the amphorae and pots of ancient shipwrecks. They have been known to escape the rectangular glass panes of aquarium tanks to slip into teapots. Their need to secure themselves within burrows strikes me as a primordial insecurity, a need to return to the protective shell that once covered their ancestors, which evolution took away before the creature inside — so shapeless, so appallingly naked — was ready to be exposed.

But they are far from defenseless. Caught in the open, they can instantaneously change color — even change the texture of their skin — to blend into the background. When startled, they eject a cloud of ink to mask their escape, like a magician disappearing behind a puff of stage smoke. As predators, they are armed with excellent eyesight, unbelievably supple, muscular bodies, and arms and sucker disks exquisitely sensitive to touch and chemical sensation. They can snatch a crab from the cover of their hole with a single arm, or they can leap up and descend upon it with their mantle flared like a malevolent parachute. Their mouth is a hard bony beak like a parrot's, and a small octopus can spurt poison from its salivary glands potent enough to kill a rabbit. When they eat a crab or a lobster they send their probing, flexible arms deep into its crevices, conveying the meat out along the sucker disks and leaving behind an almost unbroken, hollow shell.

Octopuses are solitary. They never congregate and generally can't seem to stand the sight of each other. But they mate with languorous absorption, clinging to one another for hours. The male has a specialized arm called a hectocyclus, with a long groove in the center, which he inserts into the female's body cavity. The mating itself has a kind of loading-bay quality to it — the male delivering sperm packets at regular intervals down the groove in his arm, the female receiving them and warehousing them in her oviduct.

Afterward they part company. He slithers off to business as usual. She returns to her lair to lay her eggs, brood them, and die. Many thousands of eggs are encased in long white tubes that the mother octopus hangs from the ceiling of the hole like icicles. While her young are developing, she will not eat. Her will has been streamlined to a maternal obsession so total it can end only in death. She cleans the egg cases with her arms and with blasts of air from her funnel. Though she weakens daily, she rouses herself to chase off potential predators. When the young hatch, there may be a quarter million of them, a cloud of pinhead-sized baby octopuses drifting out from the hole where their mother lies depleted and dying.

Octopuses sometimes eat their own arms; sometimes they grow them back. Their own ink is toxic to them, but they have been observed to walk through the flames of a beach-side campfire with no apparent discomfort. Octopuses are susceptible to hypnosis, and if stared at long enough will turn as limp as mops. The surest way to kill one is to bite down hard between its eyes.

Many species of octopus, some of them still probably unknown, inhabit the world's oceans. Far below in the abyss, at depths of up to three miles, live blind octopuses with translucent gossamer tissues. The blue-ringed octopus, found in the waters of Australia, has a venomous bite that has proven fatal to humans. Reports of giant octopuses tend toward the folkloric, but perhaps such creatures do exist. A partial specimen that washed ashore at St. Augustine, Florida, in 1897 supposedly had arms 75 to 90 feet long — hard to believe, especially given the fact that an octopus with arms that long would have an armtip to armtip span of 180 feet.

There is no particular reason to fear octopuses, but there have been, from time to time, "attacks" in which an octopus impulsively whipped out an arm or two and pinioned a passing bather. Once an Australian parson, leading a group of boys in a walk along the shore in Victoria, had the peculiar experience of being set upon by an octopus that crawled right out of the water to grab him around the legs and waist. (He beat it off with a fishing rod.) Certainly a surprise encounter of this sort with an octopus is an unsavory experience. "My mouth was smothered by some flabby moving horror," recounts one traumatized octopus wrestler. "The suckers felt like hot rings pulling at my skin. It was only two seconds, I suppose . . . but it seemed like a century of nausea." But anyone who has ever come across an octopus in open water, watched it spurt ink and slink deep into its hole, its eyes full of fear and its chromatic skin flashing in alarm, knows that a deliberate attack is a wildly improbable event.

I was not frightened of octopuses, but more than any other creature on the reef they haunted my imagination. And tonight, as I shone my light into the hole and stared at that octopus eye, with its eerie black rectangular pupil, I was startled by a feeling as intense as an electric shock. The octopus did not just have a brain, it had a mind. And that mind, unthinkably alien, was trained on me. A thought was passing between us. The eye looking out at me was filled with fear and disdain, and its message was as clear as if it had been spoken: Go away.

I swam off, leaving the octopus clutching its beer can, but the sea still seemed charged with watchfulness. I thought this was an illusion until I noticed, hovering alongside me, a school of squids. There were perhaps twenty or thirty of them, and more trailing off into the darkness. Because each squid swam in the same plane as its neighbor, the school appeared as a thin horizontal band of matter, like an asteroid belt drifting in space. The squids were studying me. They seemed to be waiting to discover what my next move would be. When I shone my light on them I could see the flashing interest in their eyes, but when I moved toward them they backed up in unison, then closed in again when I withdrew.

Like octopuses, squids are cephalopods — mollusks on a breakaway evolutionary track. They have the same vivid, knowing eyes, but they have ten arms instead of eight; they do not den up like octopuses but spend their lives soaring in squadron formation through the open sea. There are hundreds of species, ranging in size from less than an inch to more than sixty feet, and their alien variety strains the imagination. Some species can expel water through their siphons, rocket out of the water, and soar along above the surface for hundreds of yards like a flying fish. The eye of a giant squid, the largest eye of any creature on earth, can reach a diameter of 15 inches. Some squids that live in the deep ocean have glow-in-the-dark eyes set on the ends of waving stalks. All are hunters. Stalking their prey, they move with ghostly finesse. When they are within striking range they burst forth in a single propulsive lunge, their arms flared to entrap the victim. In most species of squids, contact is made first with two spectacularly long tentacles, heavy and flared at the end like clubs and covered with sucker disks rimmed with tiny teeth. These powerful tentacles grasp and stun the prey and deliver it to the squid's mouth, which, like the mouth of the octopus, is a beak designed to chomp and shear off hunks of flesh. Squids are ravening, unearthly beasts. Like sharks, when excited they may fall into a cannibalistic feeding frenzy, grasping and slashing at each other with chilling indifference. A Humboldt Current squid, which grows to twelve feet or so, can easily dispatch a sixty-pound yellowfin tuna. On one occasion in World War II, a group of British sailors, clinging to a life raft after the sinking of their troopship, were attacked by a large squid of an unknown species. The squid wrapped its tentacles around one of the men and dragged him down screaming into the ocean. He never reappeared. Another man was seized but released. Fifteen years later he still bore the circular scars from the squid's sucker disks.

We know of giant squids from myth, conjecture, and various biological oddments. From time to time their reeking remains are found washed up on beaches or trapped in trawling nets. The giant squid is one of the largest creatures in the ocean, but no one knows much about it, or has even had a clear look at a living specimen. It is known that giant squids "do battle," as children's picture books quaintly put it, with sperm whales deep in the ocean. The whales have sucker scars to prove it, and on more than one occasion harpooned sperm whales have been observed, in their death agony, vomiting out colossal chunks of squid. It is presumed that sperm whales routinely hunt giant squids, seizing them in the lightless depths and chomping down with their powerful jaws while the tentacles coil in fury around the whales' great locomotive-shaped heads. No one has ever witnessed this spectacle underwater, but Frank Bullen, who wrote the whaling memoir *The Cruise of the Cachalot,* had the sensational luck to see it one night on the surface.

"A very large sperm whale," Bullen later wrote,

> was locked in deadly conflict with a cuttlefish, or squid, almost as large as himself, whose interminable tentacles seemed to enlace the whole of his great body. The head of the whale especially seemed a perfect net-work of writhing arms — naturally, I suppose, for it appeared as if the whale had the tail part of the mollusc in his jaws, and, in a business-like, methodical way, was sawing through it.
>
> By the side of the black columnar head of the whale appeared the head of the great squid, as awful an object as one could well imagine even in a fevered dream. . . . The eyes were very remarkable from their size and blackness, which, contrasted with the livid whiteness of the head, made their appearance all the more striking. They were, at least, a foot in diameter, and, seen under such conditions, looked decidedly eery and hobgoblin-like.

Another great ocean spectacle, rarely seen, is squid sex. In certain species, mating occurs during mass orgies; millions of the animals congregate in a few square miles of ocean and turn the water into a twitching scene of havoc and desire. Jacques Cousteau and the crew of the *Calypso* encountered such a raft of flesh off the waters of southern California. In their fevered state the squids made no clear distinction between male and female, between mating and fighting, between sex and death. It was, wrote Cousteau, "like a giant puree of hyperactive microbes, magnified millions of times."

It lasted for days, a convulsive, delirious free-for-all, at the end of which the gravid females wafted to the bottom of the sea, laid their eggs, and surrendered their own weary bodies to death. Afterward, Cousteau wrote, "twenty million pale cadavers cover the flat sandy floor of the ocean." Twenty million dead squids, but billions more a-borning. Squids lay their eggs in giant milky bouquets that can measure forty feet across. After such a mating frenzy hundreds of acres of the sea floor will be covered with squid eggs.

The squids that were swimming along with me had a mild and whimsical look, their curious eyes shining in the darkness, their arms gathered together and tapering gracefully to a point in front of their heads. I liked them, and — in the way we humans helplessly do — I turned them into emblems of my own state of mind. I saw them as escorts, protectors. When, all at once, they disappeared, their pale bodies winking out of sight, I felt as if their approval of me had been summarily withdrawn and that the dark water was holding me up to judgment once again.

A shark suddenly reared up ahead of me, its greenish skin passing only a foot or two in front of my face. I jerked away instinctively, but noted with relief as I did so that it was only a nurse shark. My relief was caused in part by the shark's appearance. Except for the barbels that hung down from the corners of its mouth like a Fu Manchu mustache, the creature had a bookish aspect. Its eyes were tiny and dim — the eyes of a myopic professor whose spectacles had just been snatched off his face. It had a soft, sloping snout and a recessed mouth. I had evidently startled it out of its habitual resting place under a coral ledge, and now, annoyed, it swam off ahead of me, its powerful blade of a tail swatting back and forth, rippling the muscles along its back and flanks.

Nurse sharks have an exaggerated reputation for harmlessness. They can, in fact, be dangerous, especially if pestered or provoked. Their mouths are crowded with rows of small, squat teeth shaped somewhat like maple leaves, teeth designed to crunch up snail shells and the carapaces of lobsters and crabs, but also capable of clamping down on a human arm or thigh. Nevertheless, nurse sharks are not disposed to attacking humans, and I had no rational fear of this one. But I was spooked in general, thinking of the speed with which it had appeared, zooming into my sight from a blind quarter of the dark sea. When I swam to the surface and blew the salt water out of my snorkel, I felt vulnerable, subject to predation from below, so I stroked quietly back toward shore, crossing the sand flats and the turtle grass beds and emerging onto the beach in the warm night air.

It was late by then. I walked to my room, took a quick shower, and then heated up a frozen apple turnover I had bought at Herbert Sadler's grocery store. I ate the turnover, and then another, and went to bed feeling bloated and stricken with gravity. In the room there was a strange rustling noise whose origin I could not identify. I ignored it and tried to sleep, but the sound persisted, moving from place to place along the base of the walls. Finally I turned on the light and looked around. In the center of the room stood a big land crab, its claws extended, its body pivoting in blind panic as it saw me looming toward it. I stared down at the crab. Mysteriously, I broke into tears. I cried for half a minute, then abruptly stopped. There was no emotion to my crying, no cause that I could discern, just the tears themselves. Puzzled, vaguely ashamed of myself, I escorted the crab out the door with my foot and stood on the porch for a moment looking out at the ocean. The night was still, and the sound of the water sluicing against the reef was as faint as the sound of my own breath.

7

The Pain of Water

Once I spent the night in an underwater hotel. It lay at the bottom of a murky lagoon in Key Largo, Florida, and was billed as the "world's first luxury innerspace accommodation." I was there on assignment for, of all things, *House and Garden* magazine, and nowadays I boldly claim the distinction of being the first paying guest in the history of the world to check into an underwater hotel.

The hotel — Jules' Undersea Lodge, named for Jules Verne — was thirty feet below the surface and could be reached only by plunging into the water and swimming down to it, breathing along the way through a long air hose called a hookah. On the way down I was escorted by a bellman, who swam along holding my belongings in a watertight plastic suitcase. When the lodge came into view I saw that it was a large concrete bunker resting on four pillars sunk into the loose silt that shrouded the bottom of the lagoon like a heavy gas. From the outside it looked less like a luxury hotel than a bleak outpost on some stormy undersea frontier.

There was no door. We swam beneath the structure, where a diffuse light was visible, and surfaced through a pool into a room whose floor was covered in gleaming blue tile. When I climbed out of the pool, I found myself in a kind of entrance lobby, a refuge of compressed air. On either side of the lobby were two hollow cylinders containing the guest rooms and kitchen area. All in all the hotel was not much larger than a double-wide mobile home. In a former life it had been a Spartan habitat used for long-duration experiments in undersea living, but for the last ten years it had been sitting in dry dock. Now, submerged in a mangrove lagoon, the structure had been retrofitted with so many gadgets and deluxe appointments it resembled an underwater bachelor pad. The cylindrical rooms were wrapped in carpet, artificial plants were set on the nightstands, and next to a bank of electronic equipment a video library featured such titles as *Splash, 20,000 Leagues Under the Sea,* and *The Blue Lagoon.*

I liked it there. I found the hotel immediately comfortable and strangely serene. Because the air pumped down to it was pressurized, sounds were muted and voices tended not to carry farther than a few feet; one sensed a constant inrush of silence. The one permanent background noise was a sound like a waterfall that came from the venting of pressurized air out the entry pool. Over the kitchen sink, where in another kitchen there might have been a clock, hung a depth gauge.

At dinner time room service appeared, bringing lobster tails and key lime tarts. Afterward I listened to Debussy's "La Mer" on the CD player and looked out the Plexiglas porthole in my bedroom. Snappers and porkfish swam through the gloomy water, along with the drifting seed pods of mangrove trees. Through the window I saw intimations of the weather that presided over the surface. If the sun went behind a cloud, I could see it in the sudden darkness of the water, and when the shadows lingered I knew it must be raining above me. I had worried at first that I might feel closed in, but as the day came to an end the space I inhabited seemed instead to expand. Evening deepened underwater just as it does on the surface, and finally all I could see out of my window was the undifferentiated darkness of the lagoon. I don't remember dreaming that night, but just before I dropped off to sleep I imagined that the lodge was rocking slightly and that the depth gauge in the kitchen had registered another few feet.

To sleep underwater! How the child I had been would have loved the idea that he would one day be here, snoozing at the bottom of a lagoon while horseshoe crabs mated outside his window and shrimp leapt from the depths of the entry pool to perish, bewildered, on the blue-tiled floor. It was all he could have asked for — to be in this mysterious, tucked-away place, his sleeping mind roaming and merging with the water that surrounded him.

And how he would have loved to be here in Grand Turk, dawdling underwater at Harmonium Point as the day's diving draws to a close. Rising from the promontory, I make my slow way to the boat. Overhead I can see the other divers already on the surface, taking off their backpacks and handing them to Mitch as he leans over the gunwale, his bearded face clearly visible through sixty feet of ocean. When they heave themselves out of the water and over the side of the boat, it looks as if they have climbed through a mirror. I have air left, and I linger, moving slowly up the water column, feeling the increasing sunlight, the expanding air in my lungs. It is like leaving the earth in a balloon — slowly, silently ascending as the landscape spreads below me, grows vast and panoramic. I see the bone-white flats covered with staghorn coral rubble; the coral heads beneath a grazing swarm of parrotfish; the lizardfish, rays, and flounders that are nestled beneath a covering of sand. There is the dense coral bank, rounded and gradual on the shoreward side, shearing off suddenly into the black depths of Turks Passage. From above, the wall is fissured with grooves and surge channels, where schools of blue chromis meander above the white sand.

At ten feet below the surface, I make a decompression stop. There is no need to hold on to the weighted line that hangs from the boat to maintain my position. I am absolutely neutral. I point my fins downward and hug my chest with my arms. I watch a bubble of water in my mask travel around the perimeter of the lens. Off to my right are several dozen barracuda. Motionless like me, they face the current with hypnotic patience, waiting for something to excite them, their shining bodies arranged in tiers like the rungs of a ladder.

I feel disembodied and transparent, the seawater seeping through me as if I were some membranous organism like a jellyfish. Sometimes I even forget I am underwater, and my body simply accepts its unbounded freedom of movement — its spirit-lightness, its ability to glide off at will into every dimension — as a condition that has been naturally bestowed and will not be revoked. Here at my decompression stop, the barracuda hovering companionably nearby, schools of fish passing below my feet, I wonder: What if I closed my eyes and let my mind just idle down into unconsciousness? Would I still keep breathing through my regulator? Would I continue to hover with perfect trim at ten feet? Could I *sleep* here?

"Water carries us," writes the French philosopher Gaston

Bachelard. "Water rocks us. Water puts us to sleep. Water gives us back our mother."

Here beneath the ocean, a psychiatrist might say, I am searching for a sort of maternal oblivion. I want to dissolve away, not to die but to recede into the sleepy, safe, nonthinking being I used to be, before I was expelled from the womb and supplanted by my conscious self. Water is the human mind's most powerful mother symbol — a reminder of the benevolent, nurturing void from which we all emerged. Pregnant women, in their dreams, see floods and gentle waves and ocean tides, see their children borne seaward on moving sheets of water or slowly bobbing to the surface, reaching out their tiny hands for rescue. At the baptismal font — designed, it is said, to suggest the womb of Mary — the sacrament of rebirth is conducted by touching the forehead with flowing water or by immersing the entire body so that the candidate can surface newborn, blinking and shining with grace. When the Vedas speak of water, the word is *matritamah* — "the most maternal." Even the letter *M*, the universal "Mother letter," is an ideogram meant to depict the movement of waves across the surface of water.

I remember the odd satisfactions of diving in submerged caverns in central Florida, floating through dark halls of polished limestone with the beam of my light piercing the pure water like a laser. The dangerous challenge of the sport was to find an ever more distant place, far removed from surface light and air. It was a search for a secret pocket of the earth that was yours alone. And to reach such a chamber, to experience an imitation sense-memory of the perfect safety of the womb, it was necessary to risk your life.

When diving in caves, I was always cautious, adhering without question to every conceivable rule. I saw how seductive it was, how death in an underwater cave would be, for all its horror, an infinite form of regression. You would go back the way you came, swimming through a dark canal, divesting yourself of consciousness and matter until finally you were reabsorbed into the bloodstream of the earth.

It has always been the case for me that, the stranger the sensation, the stronger the conviction I have felt it before. I swam through those Florida caves in a perpetual nimbus of *déjà vu*. With every stroke of my fins, I felt closer to something I had visited once before, longer ago than I could imagine. It was unbearably tantalizing. Once, to find the opening of a cave, we had to swim out to the middle of a muddy river, sink down and feel with our hands until we had located a certain submerged log, and then tie our line to it. Underneath the log was an extraordinary sight — a hole filled with clear water leading into the bedrock, looking like a trap door to another dimension. We entered the hole, unspooling the line as we went. Under artesian pressure, the clear water shot upward through the tunnel with ferocious velocity. It was like crawling the wrong way down a vacuum tube. With one hand, I held my mask to my face so it wouldn't be ripped off, and with the other I groped for handholds in the rock to pull myself down.

The lights of the divers in front of me were as distant as starlight. When I looked above, I could see the muddy river bottom drifting like some ghastly cloud above the clear water of the flue. Finally I dropped from the pressurized tunnel into an elongated room, where the outflow was not so concentrated. We regrouped here, made our okay signs, and looked into one another's distorted eyes. The walls of the room were black, as if they had been burned, and they rose to join overhead like the vault of a cathedral. I drifted up off the floor and hung in the middle of the room, my ragged breathing suddenly quieted, my air bubbles pooling in silvery globules against the limestone ceiling. The cavern was familiar to me. I had been here before, and it had remained all this time, waiting for me to stumble upon the path that led back to it. I turned a half-somersault in the middle of the room and hung there upside down, feeling suddenly entrusted with the knowledge that there are no undiscovered places, only unremembered ones. My awareness, I understood, was a vast untapped pool. How many more rooms like this one did it hold, how many more memories waiting to be brought to light?

Psychiatrists use the term "Oceanic Feeling" to describe the sensation that steals over people when they begin to suspect that their own selves are a part of some cosmic whole. It is, says the psychoanalytic encyclopedia I consulted, a "primitive" feeling, "the source of the religious spirit." The Oceanic Feeling can be experienced anywhere, of course, but its name is certainly no accident. When you are underwater, you can feel the ocean trying to seep through the walls of your cells and flood your body with salt water. It wants to pull you down, to kill you or christen you, to make you part of its limitless self.

Sometimes while diving I have lost the distinction between my own being and the element through which I am moving. I have been swamped by the Oceanic Feeling, most often after being lulled to a dangerous state of susceptibility by nitrogen narcosis. Better known as rapture of the deep, nitrogen narcosis commonly afflicts scuba divers when they have descended below a hundred feet or so. The deeper a diver goes, the more the compressed nitrogen in his tank — and in his bloodstream — begins to act as an intoxicant. All at once he feels invincible and unaccountably mirthful. The diver offers his regulator to a grouper, and the deadpan expression on the fish's face breaks him up.

I've only been seriously narked once, and the experience was not rapturous but frightening. It happened in a Florida sinkhole known as Forty Fathom Grotto. The sinkhole was a circular basin in the bedrock no more than a hundred feet across, and its water was dark and turbid, conditions that are known to mysteriously accelerate the onset of nitrogen narcosis. A dead horse floated on the surface, trailing a slick of offal. The other two divers and I tried to ignore it, but as we suited up at the sink's edge I knew we were courting bad vibes. We dove deep, heading toward the rubble cone that rose from the bottom. At about eighty feet I began to feel detached, suddenly aware that my rational mind was no longer such a keen instrument. If I'd had sense, I would have gone back up then; but of course I had no sense. Watching the other divers go deeper in the murk, I followed, wanting above all else not to be left alone. When I try now to recall the feeling that came over me, I can only conjure up a kind of fearful absence. My conscious self blinked in and out, and in its place was a dark angel, a wraithlike, panicky form that was only vaguely connected with my own displaced awareness. Most of the time I could detect only a depersonalized dread, but every so often I would be jerked alert by the vestiges of my functioning mind. *"Wait a minute!"* I would try to shout to myself, *"Where am I? Who am I?"*

We hit bottom at 135 feet. Outwardly I appeared normal. The other divers, who had not been affected by nitrogen poisoning, saw nothing wrong. With sputtering attention I managed to notice my surroundings. Here was a car, a Ford Falcon. I learned later that it had probably been rolled into the sinkhole by someone eager to report a stolen vehicle and collect the insurance. But at the time I gave no thought to why it was there. I put my head through the driver's window and looked inside. The shredded, floating upholstery coalesced into a grisly hallucination. I found myself staring at a phantom driver, wispy and malevolent, slowly turning his head to present me with a welcoming grin.

"Get in the car," he said, "and let's go for a drive."

I wondered if I had a choice. I felt an urgent command coursing through my body: *Up!* And as I moved away from the phantom driver toward the pale light above, I could feel the scattered pieces of me — my mind, my body, my soul itself, perhaps — come together again. Even the dead horse floating on the surface seemed to be welcoming me back.

Terrifying as those moments at the bottom of the sinkhole were, they had a powerful narcotic appeal. For a time it was as if I had simply disappeared, as if my body and mind had passed through a sieve so fine that I emerged on the other side as only a dispersing haze of molecules. I came away with a memory of *not being there,* but of being alive and obscurely alert, and the memory was achingly familiar. I had known this loss of myself before, and thought that I would again. If I had drowned in the sinkhole, I thought, I would not have noticed the process very much, but on some level would have felt the satisfaction of being reabsorbed, cycled back into the water.

Even at shallow depths in the clear waters of the Grand Turk reef, I could often feel the first dreamy effects of nitrogen narcosis. When I was suspended in seawater, the difference between myself and the world around me was never as sharp as it was on the surface. My hearing was muffled, my sense of smell was apparently inoperative, and the constant pressure of the ocean on my skin limited the range of my tactile awareness. But I felt the rapture of the deep; it was the rapture of belonging. This was my territory. This was, in some sense, me. *I* was the ocean, and my body was nothing more than a particle within it, an uninhabited probe coasting through the radiant waters of the reef, witlessly receiving sense impressions.

Sometimes, even when my mind was fully functioning, I forgot I was underwater at all, and it thrilled me to realize this. The more casual the experience grew, the more ecstatic I became, because I could feel the water accepting me, letting its guard down. It was the same feeling I remembered having when I was three or four, standing before a painting in my grandparents' house, a moody autumnal scene of boats and houses crowded upon a European canal. I stood there wondering not *whether* I could enter the painting, but *how*. There was, I thought, a simple trick that lay just outside the reach of my knowledge, but any day now I would stumble across it.

Like that painting, the ocean provoked me, drew me in. I did not know if it was leading me forward or back, but I felt constantly that I was on the brink of finding out.

"A being dedicated to water," Gaston Bachelard warns, "is a being in flux. He dies every minute; something of his substance is continually falling away. . . . Water always flows, always falls, always ends in horizontal death. . . . The pain of water is infinite."

8

The Dead Whale

> The weather started getting rough,
> The little boat was tossed.
> If not for the captain and his fearless crew,
> The *Minnow* would be lost . . .
> . . . .
> Oh, the *Minnow* would be lost.

Over the roar of the outboard, Mitch belted out the theme song to "Gilligan's Island." He stood up in the stern, steering the boat by moving the outboard motor handle with the calf of his leg. He wore a baseball cap and sunglasses held in place by a Croakee, but the usual brilliance of the sun was muted today, the horizon crowded with ranks of gray cloud.

Weather and wind were starting to come in from the northeast. In town the streets were dappled with white blossoms that had been shaken loose from the trees, and the water on the lee side of the island had begun to stir. The underwater topography, normally denoted by the surface hues of the sea — the turquoise sand flats, the dark green patch reefs, the azure depths of the wall — was lost now under a uniform blanket of blue. The water visibility was poor, and we could see the coral only when we were right on top of it — a pale ocher smudge below the blue chop, like a deposit of slag. Up ahead, something big stirred the surface of the water and then disappeared.

"Did you ever see a humpback out here?" I asked Mitch as I gazed out at the water, hoping the mysterious creature would surface again.

"Only once on scuba," he said. "I was on the south end of the island one day, diving out of a thirteen-foot Whaler. As soon as we started the dive we could hear them singing. Every few minutes they'd let out this long, sustained kind of moan. It got louder as they got closer to the boat. We kept heading out from the wall, looking for them. Finally I looked out and saw this, like, fifteen-foot-long white shape. I thought, that's small for a whale. Then I realized it was just his pectoral fin. One of them swam right by me, except it wasn't even like he was swimming. It was like a semi driving by real slow."

Mitch bent down to adjust the throttle on the outboard, idling to a stop at a dive site several hundred yards offshore from the chief minister's house, known locally as Cocaine Towers. At the moment, the house was without an official occupant. The former chief minister, Norman Saunders, was in jail in the United States on drug-smuggling charges. His arrest and conviction had been a crippling blow to the nascent tourist industry of the Turks and Caicos, and the ensuing crisis had moved the Crown to send over a political swat team from England to supervise the writing of a new constitution and to hold elections for a new local government.

On the boat with Mitch and me that day were two married couples and a solo traveler from Toronto with a sour disposition, whose conversational skills consisted solely of a running putdown of his surroundings. "My hotel," the Canadian said, pulling on his bright yellow booties and matching fins, "is a typical second-class hotel in the third world. I'd say the diving here is no better than average, nothing like Cayman or the South Pacific. And the island itself, let's be frank, is pretty dreadful." He strapped on his $600 dive computer, picked up his underwater housing and strobe bar, and, after uttering a few last-minute complaints, did a backward roll off the boat.

A couple from Colorado, experienced divers in matching custom-made wetsuits, followed him overboard. Next were Patty and Larry, both employees of a big hospital corporation in Los Angeles. She wore a leopard-print bikini, he a blue Lycra skin suit. Patty said she had done quite a bit of diving in the past but had been ill lately with a bone infection, and it was only in the last few days that she had been disconnected from an IV that dripped penicillin into her veins. This was Larry's first open-water dive, and when he rolled off the boat he made the common mistake of forgetting to hold his mask to his face. When he hit the water, the mask came off, leaving him disoriented and struggling on the surface.

Patty and I hovered near him until he had pulled his mask back on, and then followed him down as he made his fitful way along the down line, stopping every few feet to waggle his head and try to clear his ears. Ear squeeze is often a curse to beginning divers, causing not just pain but incipient feelings of panic and inadequacy. The astronomer Edmund Halley, who was also a leading designer of diving bells, summed up the problem neatly in 1716: "A pressure begins to be felt on each ear, which by degrees grows painful as if a quill were forcibly thrust into the hole of the ear; til at length, the force overcoming the obstacle, that which constringes these pores yields to the pressure, and letting some condensed air slip in, present ease ensues."

If Halley's "present ease" does not arrive, the result can be a ruptured eardrum. Ear squeeze is caused by a pressure imbalance between the middle ear and the surrounding water — a situation that is easily correctable given enough practice and conditioning. It had been years since I was afflicted by ear squeeze, having finally stumbled upon what was, for me, a foolproof method of clearing my Eustachian tubes. (For the record, here it is: an hour before diving, lean your head to one side, pinch your nostrils with your fingers, and blow softly through your nose until you hear a squeak in the opposite ear. Then lean your head in the other direction and repeat the process. Do this every five minutes or so, and continue to do it once you enter the water and periodically as you descend.) But the pain and frustration of not being able to clear my ears, of having to proceed cautiously and incrementally downward when every nerve in my body told me to zoom heedlessly to the bottom, was still a vivid memory of spoiled fun. Larry's ear squeeze aroused my sympathy and my protective instincts.

The squirrelly conditions did not improve. As the sea kept building on the surface, the visibility faded below. We could see about twenty feet in any direction, a far cry from the 200-foot-visibility days of high summer that Mitch had been telling me about before the wind came up. Swirling particles of sand gusted along the lip of the wall, and in the face of the oncoming storm I thought I could detect a charged, excitable mood among the creatures of the reef. A school of blue chromis angled down in front of me, filtering deep into a meandering coral channel as if searching for shelter. When I stopped swimming for a moment to take notes, a cloud of yellowtail snappers appeared and began to nibble excitedly at the edges of my slate. In growths of soft coral, trumpetfish were waiting in ambush. Thin as knife blades, they hung upside down in perfect mimicry of their surroundings, apparently just more rippling gorgonian stalks. Small fish passing beneath them were in mortal danger. In the sea, attacks almost always come from below or from the side, but the trumpetfish, hanging head down, seizes its prey from above, slurping it up into its mouth faster than the eye can follow.

I watched a trumpetfish back slowly out of a soft coral formation next to me, its hunting spoiled by an aggressive damselfish a tenth its size fearlessly defending its little plot of algae. The trumpetfish assumed a horizontal position and swam ahead, looking for another place to hide. The fish was two and a half feet long; it moved slowly with the unhurried certainty of a predator, its head angled down, its long straight shaft of a body powered by a fluttering trio of fins that reminded me of the feathers on the end of an arrow. In the foggy underwater illumination the fish's skin, with its silvery stripes, was an ever-variable shade of brown. At moments, when it caught the scant sunlight just right, it had the glint of brass. Bony and narrow, the fish looked less like a trumpet to me than like some sort of specialized implement you'd buy at a hardware store for pulling up hard-to-reach nails.

I heard my name being called. It sounded as if it came from a mile away, a constricted groan so low and slow I could almost see the sound waves batting against the water to reach me. I heard it again — "Stuuuuuuuuuueve" — and looked around to see Mitch hovering at the wall, holding his regulator in one hand and cupping his mouth with the other, shouting my name underwater in a violent expulsion of breath.

When I swam over to him I saw that he wanted to show me something. In his hand he held a greenish, floppy object six inches long, flat as a tortilla, its dorsal surface crowded with fleshy gills that looked like suction cups. It was a nudibranch, specifically a lettuce sea slug. Nudibranch means "naked gills." In other mollusks the breathing organs are hidden in the folds of their mantles, but nudibranchs have them out on flagrant display. Not just the gills are naked. A nudibranch, essentially, is a snail without a shell, a shapeless, wriggling creature that seems not only defenseless but luridly exposed. Sea slugs are not defenseless, however. To discourage predators, they secrete a repellent slime. And though they have no natural armament of their own, they eat coral polyps and other creatures that are loaded with stinging cells. Sea slugs have the startling ability to save these cells from digestion and put them to use in the defense of their own bodies.

Mitch handed me the lettuce slug, and I let it undulate across my palm and fingers. It felt like wet pasta. Like so many other things in the ocean it was beguiling and repugnant at the same time. I remembered how, diving one night off Heron Island on Australia's Great Barrier Reef, I had come across a Spanish dancer as it made its billowing way through the water. Like the lettuce slug, a Spanish dancer is a nudibranch, but it is huge and florid. The one I saw was the size and shape of a serving platter, and in the beam of my underwater light it was a brilliant, sumptuous shade of red. Every motion the Spanish dancer made seemed to pump more redness into it, so that its vivid color had a dynamic quality, like a rising blush. The creature wafted through the water, flamboyant and ghostly. One of the divers reached out her hand and stroked it, as if it were a thing that needed comfort. I touched it too. It was slick and satiny, and though it was light as a veil I could feel its strange organic heft. After a while four or five divers had gathered around it, reaching out for it. In the cold ocean night it shone like a flame and seemed to warm us with its beauty.

Mitch showed the lettuce sea slug to the rest of the divers, and then we headed south along the wall, the visibility worsening. Ahead of me I could see the disembodied yellow fins of the malcontent diver from Toronto as his feet moved rhythmically up and down in the haze. In a moment even they were no longer in sight. A kind of underwater sandstorm came up, obscuring everything, turning the once-translucent water into a slurry of drifting particles. Both Patty and Larry, with whom I had been swimming, immediately disappeared in the fog, and it took me a few moments to find them again. Larry was spinning in slow circles at sixty feet, looking for his wife. I made a sign for him to stay put and then swam down to eighty feet, where I had last seen Patty. She was still there, her leopard print bikini drifting into view in the murky water. Her mask was half flooded and her eyes were wide with concern. "Where's Larry?" she wrote on her slate in a loopy, concerned hand. I pointed up, took her elbow as a kindly old gentleman would, and ushered her up the wall to where her husband was waiting.

Under way again, I spotted a surge channel and banked left, leading Larry and Patty up through this crevice to its exit near the top of the wall. Then we turned south, back toward the boat. I was nervous; all the landmarks were obscured by a blowing curtain of sand and silt. I didn't want to overshoot the boat, to surface with two frazzled divers in a rough sea and pelting rain. I felt that we were in no danger, but I was also aware from years of diving that there is no real way to gauge danger underwater. Being underwater is inherently perilous; benign conditions do not exist. Safety is a matter of finding comfort, of feeling secure in the quality of your responses — of feeling at home. When your confidence craters, when you feel suddenly diffident and unsure, the ocean can turn as hostile as a snarling dog.

The ocean was not hostile that day, just inconvenient, but I was too much of a worrywart to feel comfortable looking out for anyone but myself. When Larry's gauge started to nose toward the red, and my dead reckoning told me we were reasonably close to the boat, I signaled for us to surface. At forty feet I heard a dramatic pounding noise coming from above. I thought at first it was the propeller of a boat or somebody's free-flowing regulator, but as we cautiously made our way to the surface I saw it was a wind-driven downpour, the rain hitting the waves with the concerted violence of a waterfall. When I slipped my head into the storm I could see the boat waiting amiably for us, Mitch already on board and helping the other divers with their tanks and fins.

"Who is the tall dark stranger there?" he was singing above the pounding rain. "Maverick is the name."

Another spell of rain and variable, infuriating weather. Mitch took his boat to its anchorage in the protected sound on the south end of the island and left it there for a week, then took to his bed with an ear infection brought on, he said, by a case of "reverse squeeze," an inability to clear his ears while ascending to the surface. The wind came from the east, though sometimes for an hour or two in the morning it was calm and the sun was out — not only out but fierce. It hammered down the ripples in the sea, making the surface as white and flat as a playa lake in the desert. But as soon as I had my gear gathered together the wind would build again and the sun would disappear, taking with it the promise of calm, clear water.

In the late morning I stood on Front Street, gazing wistfully out at the buoy that marked Harmonium Point, and in an agitated frame of mind mourned the loss of these precious diving days. I decided to console myself with lunch, and walked up the road to Peanuts' restaurant, a corrugated tin shack painted blue, across the street from the Odd Fellows Lodge. Peanuts was out in front, scrubbing away at a hand-painted menu with a toothbrush and Colgate toothpaste. She explained that the sign painter had written in the wrong price for a bottle of malt — seventy-five cents instead of a dollar.

"Come on inside," she said, "and tell Peanuts what you want to eat."

I followed her through a tiny front room that contained a nonfunctioning jukebox, a refrigerator full of beer and soft drinks, an ad for Nestle's condensed milk, and glass jars full of Mars bars and penny candy. There was also a folded-up cot, where she slept at night. She was an old woman who had come to Grand Turk as a girl from North Caicos, and those were about the only two places she had ever been.

"I seen you before," she said. "You be walkin' up and down the street from time to time?"

"Yes, ma'am," I said, "that's me."

In the kitchen she had me look into a basin filled with salty water and chicken parts. She told me to point to the two pieces of chicken I wanted, and when I had made my choice she took them out and sprinkled them with seasoned salt, Shake 'n Bake, and a few other spices that she shook out of unlabeled containers.

"I made stewed conch last night," she said, patting down

the chicken. "Not like the conch chowder they serve at those hotels. Did you know they put milk in it? Right in the stew!"

She transferred the chicken pieces to a pressure cooker, then set it on the stove and held a kitchen match to the burner beneath it. A ring of blue flame exploded into being with a loud *whomp*.

"You gots to cover the chicken," she said, setting the lid on the pot. "That's where people make their mistake. It got to cook *inside*. I take a precaution, though. I take that knob off the pressure cooker. I'm afraid of the pressure, you see. It gets too much pressure, that knob shoot up and knock your head off your body."

I sat down on a broken plastic chair and the two of us stared at the pressure cooker, waiting for the chicken to cook. Outside the kitchen door was a kind of patio with a single picnic table and a fence made of packing crates. On the other side was the water. The sea was rising. White jets of foam leaped over the makeshift fence.

"That's unusual," Peanuts said, watching the building waves. "The water hardly ever com onto the earth any more."

When the chicken was done, I sat and ate it on the rickety kitchen chair while she stood over me, watching. I sipped a bottle of malt, recoiling at the thick molasses taste. I asked her how she planned to vote in the upcoming elections.

"I don't pay no attention to that. No attention. I want it to be like it was. Used to be we had the navy, we had Pan Am. Long time ago we had the Seabees."

She looked out the kitchen door at the swollen sea threatening her patio.

"It were gorgeous here then," she said.

That night I heard that a whale had washed up on the windward side of the island. Two Turks Islanders were talking about it in the bar at the Island Reef. One was saying they should get a boat and tow the whale over the reef to the other side of the island. There, if a doctor certified the meat as safe, they could begin selling it in the fish market.

"Maybe," his friend said. "I wouldn't eat none of it though. They say a whale has so much oil, if you sit down to eat it, by the time you get up the seat of your pants be wet."

I got the underwater light out of my gear bag and went walking down the beach in search of the whale. There was no moon, and the batteries in the flashlight were weak. Tiny ghost crabs flitted ahead of me on the sand, while their larger cousins wheeled out of my path with their arms extended, as if ceremoniously making way for my approach. A sandpiper kept taking off and landing, trying for some reason to keep within the pale light of the flashlight beam.

I walked for a mile in the darkness, expecting to locate the whale by the smell of its rotting carcass. But I could detect nothing, and finally decided to go home and return in the morning. The batteries were almost gone by now, so I turned the light off to save them. Even in the dark, the ghost crabs were apparent, as pale and fleet as meteors in a hazy sky. Walking along, I smacked my shin against a fishing boat and sat down for a moment in the sand to feel sorry for myself. Turning the flashlight on, I saw a big stingray near my feet, lying on its back in the sand. Its wings had been sliced off, and the line it had been caught on still hung out of its mouth, along with a pulsing quantity of maggots. Nearby, nailed to a piling like some bizarre ornament, was the desiccated body of a flying fish.

In the morning I retraced my steps in the company of Arthur Lightbourne, the twenty-five-year-old manager of the Island Reef. Arthur was born on Grand Turk, the grandson of a turtleman, but he had spent enough time in Miami and the DR to develop urbane tastes in leisure wear and stereo equipment. Though he was savvy about the first world, he still regarded the tourists who came to Grand Turk with an islander's native bemusement.

"Tell me this," he said to me as we walked down the beach. "Why is it soch a big deal for the divers to see a monta ray?"

"I don't know," I said. "They're big. They're pretty."

"Or sharks? I myself om more scared of the barracuda than the shark. Barracudas are fast. You con be in the water five feet from shore, the barracuda he con be a hondred yard out. If he want to cot you there nothing you con do. The barracuda, Steve, he don't cot you with his teeth, he cot you with his speed."

We saw the whale in the distance, like a miniature headland jutting out from the curve of the shore. There was no smell; the wind was with us. When we got closer we saw that the whale had been dead for awhile and that sharks had worked it over. It had not come completely onto the shore but was dragging bottom a few yards out. It was a mess, and for a long moment I could not tell which end was which. What I took to have been its head was a shredded, sopping mass of flesh. The whale's intestines, inflated with gas until they seemed ready to burst, floated out from the body in pale blue coils. Its broad tail, almost severed, flopped from side to side in the oncoming waves, fouled like a propeller in the turtle grass.

"That's a small tail, Steve," Arthur said, "to drive soch a big fish. Look at that. The shark took his flipper off with one bite. Then after he over the reef, the little sharks they bite off his tail. Boy, you bet there be one big bunch of hoppy fish out there now. That's gonna bait all those great white sharks. When the tide's high, they'll cross the reef and com get this boy."

Arthur said it was a humpback, but I didn't think so. For one thing, it was too early for the humpbacks to arrive from their feeding grounds in the North Atlantic to the warm Caribbean shoals where they would mate or give birth. And though this whale had been imposing when I first saw it, I realized now it was no longer than twenty feet, half the size of a mature humpback. Finally, it lacked even a remnant of the humpback whale's most conspicuous field marks, the long white pectoral fins, graceful as angels' wings, that are sometimes a third the length of the creature's body.

I waded out into the water for a closer look at the whale, sidestepping the floating entrails. The east wind that had washed the animal ashore was still strong, and the carcass scudded and teetered in the waves. It seemed to be lying on its side, and the exposed underside of its body was as wrinkled and saggy as an underinflated air mattress. When I waded around to the other side I ran into a wall of putrid air. Holding my breath, I looked for the dorsal fin but saw only a chewed-down ridge of flesh, the half-moon imprint of sharks' bites still vividly apparent in the blubber.

I wanted it to be a humpback. For all the diving I'd done, for all the years I'd lived on the ocean, I had never seen a great whale, except for a dead baby sperm whale that had once been on exhibit on a bed of dry ice in the parking lot of a Corpus Christi shopping center. The sperm whale, I remembered, had been a pale deathly gray, as gray as a fungus, and its skin had been covered with welts and freeze-dried scars. A whaling scene was painted on the wall behind it, but the baby whale seemed to bear no relation to the creatures depicted. Lying there in repose, shut tightly into its own death, it was — for reasons I couldn't grasp — a fearful thing to me, something that had leaped out of the unconscious mind and then been brutalized and beaten into stillness.

But the sight of this whale on the beach did not disturb me. Its rotting bulk was not offensive. When Arthur went back to the Island Reef, I sat alone on the beach taking snapshots of it and wondering what this ruined and shapeless thing had been. It was not a humpback, but I let my mind wander anyway to those majestic whales (as graceful, the whaler and naturalist Charles M. Scammon wrote, "as a swallow on the wing") that would soon begin to make their way south from the cold waters of Newfoundland and Iceland. I would not be here when they arrived, two or three months from now, and the thought was hard to bear.

In the North Atlantic they had spent the summer gulping down small schooling fish and massive quantities of the tiny shrimplike crustaceans known as krill, each adult whale consuming over a ton of food a day. When they began to migrate south, they would not eat at all. Humpbacks are rorquals, a word that derives from a Norwegian phrase meaning, roughly, "furrowed whale." On the undersides of their bodies, rorquals (which include the largest animal of all time, the blue whale) are outfitted with longitudinal pleats that expand like a bellows, increasing the whales' food-filtration capacity. Rorquals are baleen whales; they have no teeth, just horny, fibrous plates hanging from their upper jaws. They feed by opening their mouths and taking in gallons of seawater, then using their powerful tongues to force the water back out through the baleen sieve, leaving behind a nutritious residue of krill, fish, mollusks, jellyfish, and sometimes even unfortunate seabirds.

Some baleen whales collect their food by merely skimming along on the surface with their mouths open. Some plow into the ocean bottom with their heads and gulp down clouds of sediment teeming with dislodged crustaceans and benthic worms. Humpbacks often sweep upon their prey from below, surging upward with powerful strokes of their tail flukes and then erupting through the surface with their strange mouths agape — their throats distended with seawater, the bristly baleen plates hanging down from the upper jaws like the scrubbers on a car wash. Humpbacks, in a behavior unique among whales, sometimes dive deep into the water below a concentration of krill or small fish and then spiral upward, all the while exhaling through their blowholes. This creates a rising net of bubbles that confuses the prey and corrals them into a dense column.

Sometimes humpbacks herd fish and krill by smacking the water with their remarkable pectoral flippers. Far longer than the stubby fins of most whales, a humpback's flippers are knobby at the edges and snowy white on the bottom. Humpback whales constitute a genus, *Megaptera,* meaning "big winged." Traveling through the water, a humpback uses its wings like a bird in flight, banking on them or sweeping them back against its body to pick up speed. The flippers are flexible enough for a mother to cradle a young calf with them, or for an amorous whale to deliver — as Charles Scammon described them — "love pats," which are audible for miles. When humpback whales mate, they often do so face to face, "standing" upright in the water and grasping each other's bodies with their beautiful white fins.

"It is quite possible that humpbacks itch," write Lois King Winn and Howard E. Winn in their book *Wings in the Sea.* The whales are covered with all sorts of skin bumps, tubercles, barnacles, pycnogonids, crustaceans, and burrowing worms. Their blubber is scarred from the effects of parasites and from the nibbling attacks of such fish as the "cookie cutter" shark. Their breath, especially during the feeding season, when their baleen bristles are filled with decomposing seafood, is noxious. Humpbacks are slower than most rorquals and not as sleek, but they have a tendency to be acrobatic, roaring up out of the sea, throwing themselves onto their backs with concussive force, on occasion even clearing the water entirely. Like other whales, they engage in "spy-hopping." Supported by their powerful tails, they rear up vertically until their eyes are just above the waterline and remain there for half a minute, sometimes turning in circles, presumably to take some sort of visual bearing.

Often they loll about on the surface, their immense flippers waving back and forth. Once a helicopter from Cousteau's research vessel *Calypso* flew over a humpback whale at low altitude in the Bahamas. "Not only did the whale not sound to escape the noisy contraption overhead," Cousteau writes, "but it rolled over on its back and exposed its belly to the cool breeze created by the spinning rotor blades. It swam about unconcernedly beneath the aircraft for quite some time, slapping one flipper and then another on the ocean surface."

We do not know what whales are thinking at such moments, or how they think, or how their world appears to them. Like all cetaceans (the order of mammals that includes whales and dolphins), humpbacks have large and intriguingly complex brains bundled into their streamlined skulls. We know from the configuration of these brains, with their densely furrowed neocortices, that whales must be exquisitely alert, but the exact timbre of their intelligence will probably always be a mystery to us. They perceive their environment through well-developed eyes, a surprisingly sensitive skin, and a sparse stubble of hairs on their upper and lower jaws that presumably give them a whiskery capacity to sense nearby movements of prey. Dolphins and many of the toothed whales obtain much of their sensory information by sending and receiving high-frequency sound waves, but humpbacks do not appear to have any special capacity for echolocation. Their hearing, however, is very good, even though their external ears — tiny dimples in a wall of blubber — are almost unnoticeable.

The sounds humpbacks emit are widely construed as songs. Their booming, creaking, lowing sounds reverberate through the ocean, pass through the hulls of ships, and enter the dreams of restless sailors. "Song" is probably the wrong word to describe these mysterious emanations, but the calls have a lugubrious musical quality, and they are performed in sequenced patterns of phrases and themes that differ subtly from year to year. The songs of Pacific humpbacks are distinguishable from those in the Atlantic, giving rise to the idea that the animals have "dialects." But all the whales use essentially the same sonic vocabulary. Lois and Howard Winn have catalogued seven basic types of humpback sounds: "moans, cries, chirps, yups, oos, surface ratchets, and snores."

Humpback whale songs have been described, with some seriousness, as ballads, arias, Homeric-style sagas relating the ancestral history of the species, and heartbroken warnings about the impending nuclear annihilation of the planet. Many a hippie has fallen into a cosmic sleep to the song of a humpback whale on the turntable. When the *Voyager* spacecraft were launched on their tours of the galaxy in 1977, humpback recordings were included in the payloads on the assumption that their plangent tones would impress any aliens that might hear them in deep space. Whale songs, we cannot help but believe, are full of secret, mournful wisdom. We do not think of these songs in terms of their function but in terms of their meaning. We need to feel that they are directed toward us.

In fact, humpback songs seem to be, more than anything else, highly sophisticated and variable mating calls. Only males have been heard to sing, and only in the winter months, when the year's feeding is over and the whales have arrived at their seasonal breeding and calving grounds. The great majority of the North Atlantic humpbacks, perhaps as many as 90 percent, spend the winter in the vast midsea shoals — the Navidad, Mouchoir, and Silver banks — that fan out southward from the Turks and Caicos. A reasonable guess at the number of North Atlantic humpbacks is 5,000. The world population is at least twice that, less than a tenth the number that are thought to have roamed the world's oceans before the advent of commercial whaling. Compared to most whales, humpbacks were easy to catch. They were slow and they migrated close to land, making them vulnerable to forays from shore. Thousands were slaughtered every year — and turned into margarine. Because dead humpbacks sink, they had to be inflated with compressed air, sometimes while still alive. New Zealand whalers caught humpbacks with hollow harpoons connected to air compressors. As the whales thrashed about in agony, they were inflated. Finally the whalers killed them with a long lance that had an explosive charge of gelignite in its hollow iron point.

I came back to see the whale the next day, but somebody had towed it away.

"Excuse me, have you seen a whale?" said a woman who was walking on the beach. She was a black woman from Bermuda, prosperously dressed, and she spoke with a crackling British accent. When I explained to her that the whale had apparently been towed away, she looked not disappointed but annoyed. Her maid was with her, a girl of sixteen or seventeen named Celie who did not speak English and did not seem to understand that the whale was gone. She shaded her eyes with her hand and kept scanning the waves.

The woman explained to me that she had just moved to Grand Turk. She and her husband, a contractor, were building a house on the limestone bluffs just above us. I asked her how she liked the island so far.

"Well, not at all, really. Coming from Bermuda, you see. Perhaps you can tell me — what do people *do* here?"

I told her that people seemed to play a lot of darts.

"Fish? Fish?" called Celie, jumping up and down in the sand and pointing out to the water, where a soft mossy chunk of something was floating near the shoreline. It was about three feet across and dark brown.

"What is that *thing*?" the woman asked.

"I'm not sure," I said. "I think it might be the whale's lung."

The three of us stared at the floating object for quite some time.

"Well!" the woman finally said. "We must go now. Celie must get to work, you see."

They climbed up the bluff to their house-in-progress. I stared at the floating thing a while longer, silently speculating on what it could be. Then I walked back up the beach, got on my scooter, and drove into town. At the Poop Deck restaurant I ate a mound of peas and rice topped with slices of fried plantain. As I was leaving, I encountered an old man walking a bicycle along the salina. The handlebar basket was full of three-inch ivory teeth that he said had been pried out of the jaw of the whale. They were for sale, he said, for eighty dollars apiece. I offered him five, and he readily agreed, letting me take my pick.

The one I chose was not the largest, but it was well formed. Most of the tooth consisted of a single deep root, its surface slightly rough and discolored. Out of one end rose a smooth, conical crown, its surface as glossy as that of a pool ball. The tooth confirmed that the whale had not been a humpback, since humpbacks have no teeth. I learned later, after consulting over the phone with a biologist, that it had probably been a short-finned pilot whale, one of the small, black, bulbous-headed whales that jump through hoops in oceanariums.

Not wanting to go home yet, I walked over to the edge of the salina and sat down near the old boiling pots. The smell was pleasantly rank. An osprey soared overhead, and from the center of the island a plume of gray smoke rose from the trash being burned at the dump. I rubbed the whale tooth between my palms. Its cool, curved surface fit the contours of my hand like a fine knife.

9

Algal Ridge

One evening a few days later, while the weather was still unstable, I drove my scooter up to the Smithsonian Institution research station at the summit of the ridge that ran along the east side of the island. Inside, about twenty people had gathered for dinner. Most of them were Smithsonian employees, working on a project to raise commercially viable spider crabs by feeding them cultivated algae. The rest were either Grand Turk residents, vacationers, or castaways. Mitch was there, playing his guitar. He still suffered from reverse squeeze, and every so often he would stop strumming and painfully waggle his head.

Purple sea fans had been tacked to the walls of the Quonset hut, and the bookshelves were full of swollen, water-damaged paperbacks and stacks of scientific journals. A little boy scooted across the floor, rolling a toy diplodocus that had wheels concealed in its feet. His father, a condominium manager and shark fisherman, was telling me about the time he had been diving on the north end of the island when a nine-foot lemon shark swam straight at him. He pulled himself out of the water onto a pinnacle of abrasive elkhorn coral and waited until the shark lost interest. The next day he went out with his shark rig and caught the very fish that had chased him. "So the way it worked out was I ate *him* instead."

Rain began pounding down on the roof of the quonset hut, which reverberated like a hide drum. Two women, doctors from the vicinity of Moose Factory, Ontario, stood at the screen door and watched the rain drench the conch-lined sidewalk outside. Their vacation had been spoiled by the weather, and on the way down to Grand Turk the airline had lost their luggage, which included the parkas and mukluks they would need when they returned home.

Tomorrow they were scheduled to fly out. "We'll get into Moose Factory at midnight," one of the doctors told me. "We live on a tiny island in the middle of James Bay. If the ice on the bay is frozen enough we can take a taxi to the island. Otherwise, we'll have to hire a chopper."

"We could walk," suggested the second doctor.

"At midnight? Across the bay? I don't know, I think I'd rather stay at the Polar Bear Inn in Moose Factory."

They continued to stare forlornly through the screen door, as if expecting that at the last moment the bad weather would break and their missing luggage would be found.

In front of the quonset hut, barely visible now in the darkness and the lashing rain, stood the remains of an old telemetry antenna, left over from the days when this research facility had been a NASA tracking station. American space flights had been monitored here from the early days of the Mercury program to the Apollo moon landings. When John Glenn, returning to earth from his orbital flight in 1962, splashed down in these waters, he was hauled out of the ocean and brought to Grand Turk, where he was greeted by Vice President Lyndon Johnson. A few days later Glenn went spearfishing.

The Smithsonian had converted the base of the antenna into a kind of observation platform. Kurt Buchholz, the on-site director of the spider crab project, had taken me up there when I first dropped in at the station on a clear day several weeks earlier. From the platform we could see the whole island — the glistening salinas, the white walls and tin roofs of Cockburn Town, the boats moored in the pale blue water of North Sound. The island's windward lagoon, transparent and studded with patch reefs, stretched out below us, disappearing into the fringing white water of the reef line. Beyond the reef were isolated cays and distant islands muffled in clouds.

From this height the water in the lagoon had looked calm, but when I went out with the crew to check on the welfare of their crabs, the surface was choppy and restless. The patch reefs were extensive, and sometimes they joined together to form unbroken coral hazards just beneath the surface, so the boats had to be piloted carefully from one winding channel to the next. In some places the coral broke through the waterline, forming "boilers" of surging and sluicing foam.

The crabs were housed in a series of submerged boxes attached to floating lines. Inside the boxes were racks of mesh screens covered with algae — each screen a pastureland teeming with tiny grazing crabs. When fully grown, a spider crab is a spindly, ramshackle creature with a four-inch carapace that looks as stout as a manhole cover. I had seen them often while diving along the wall. They preferred deep water and tended to huddle inside crevices, restlessly shuffling their long jointed limbs. Spider crabs look ferocious, but they are the most delicate of herbivores. Unlike other crabs, whose meaty claws are powerful enough to crush walnuts, a spider crab's claws are thin and tapered, and the grooved pincers are made not for smashing or grasping but for plucking algae off rocks.

It takes two years for a spider crab to reach maturity. Many of the specimens in the Smithsonian's floating boxes were only days old, the size of mites and vulnerable to attacks from almost anything, including minute planktonic organisms. Others were about the size of cockroaches and had festooned themselves with a slick camouflage covering of algae.

Full-grown spider crabs are good to eat — they're similar to Alaskan king crabs — but their secretive deep-water habits have protected them from extensive human predation. The purpose of the Smithsonian project was to determine if the crabs could be raised to a marketable size using the primitive algae that accumulated on the submerged screens. Buchholz called this "turf algae" and likened it to a grassland. However, unless the screens were scraped down — "mowed" — every few days, these grasslands would turn into wilderness, sprouting growths of larger and more complex algae until the crabs' pastoral paradise was lost in a tangle of thick seaweed.

That was what we were doing today, hauling the boxes out of the unruly water and monitoring the algae growth. Thumbnail-sized crabs swarmed over the screens. With their hard carapaces they looked as indestructible as miniature tanks, but when I held one in the palm of my hand I was struck by how weightless and ghostly it felt, its tiny scrambling legs barely leaving a trace of sensation on my skin. This crab's chances of surviving to adulthood were faint. So far all of the crabs that the Smithsonian had raised had died before 200 days. Buchholz knew of only one commercially raised spider crab that had lived past that point. A man on the nearby island of Providenciales had nursed one to adulthood but had then promptly eaten it without thinking to take measurements or record other vital data that might have helped explain why the crab had survived. What was consistently striking down the Smithsonian crabs in their prime was still a matter of speculation. Perhaps the microalgae they were being fed lacked some vital nutrient that was available to them in the wild. A more intriguing theory held that the crabs, imprisoned in their swaying ocean boxes, succumbed to seasickness.

As the rain beat down on the research station, I thought of those crabs out there, clinging to their algae screens in a storm-tossed sea. The thought aroused memories of my own bouts of seasickness. Once I had spent three days on a shrimp boat in the Gulf of Mexico in a state of perpetual queasiness, rallying myself every now and then to read a chapter of *The Magic Mountain* by Thomas Mann — which, with its dense print and feverish plot developments, is probably the worst book to try to read when you're sick to your stomach — or to help the crew pop the heads of the shrimp that were dumped out of the nets onto the deck in a twitching pile.

I had been sick often on dive boats as well, in the Gulf and in the Pacific and in the Coral Sea off Australia's Queensland coast. For me, seasickness is a condition like despair, a chronic, helpless, implacable unease, as if the planet itself is trying to make my life intolerable and slough me off. At such moments the only refuge is underwater. Dangerously nauseated, overcome with diesel fumes and the motion of the pitching boat, I plug my regulator into my mouth, wait until the stern is high on the crest of a swell and leap into the smooth water of the trough before the boat comes crashing down again. Ten feet below the surface, the movement of the water is soothing and rhythmic, and I feel as if someone had just opened a door and pulled me in out of a storm.

After an hour or so, the drumming of the rain stopped. I said goodbye to the Moose Factory doctors, wishing them a pleasant return trip after their frustrating vacation, and rode my scooter down the ridge road, past the white cemetery vaults, now washed by the rain and shining in the tentative moonlight.

When I pulled into the parking lot of the Island Reef I was waylaid by a grumpy fisherman from North Carolina who had come down here, he said, to scout out locations for a commercial fishing business. I had encountered him a few times before. We had not so much as exchanged names, but he seemed to have the idea that we were long acquainted, and whenever I crossed his path he hailed me as if we were resuming an ongoing conversation.

"What about the women here?" he said as I turned off the scooter. "You got any sense of how it works?"

"Not really," I said.

"Shit! I can't seem to crack the code, you know?" He took a long dyspeptic gulp of his beer. The wind flattened one side of his gnarly hair. He was thin, with a bulging pot belly, and perhaps because of it he carried himself like a pregnant woman, with his feet apart and his shoulders back.

"I'm going to write a story for *Life* about the Haitian elections," he said, trailing me back to my room. "If that doesn't make me rich I'm planning on growing and shipping vegetables from the DR in container ships. Or a conch fishery. Jesus Christ, have you seen all the conchs just sitting around out here waiting to be caught?"

"I haven't seen *that* many conchs," I told him.

"Well, groupers then. Of course they don't have the equipment for it here. You need steel lines and winches. Grouper takes a regular hook, he runs and hides in a rock. But the minute he touches one of mine, with that steel line and that power winch, he's on his way up."

"Well, I think I'll turn in," I said when I reached my room.

"Groupers are weird fish. They're males for about three years, then they turn female."

He winked at me, as if he were describing some exotic trick performed in the brothels of the DR.

"How much do you think *Skin Diver* would pay for a story about groupers?" he asked. "Divers are interested in groupers, right?"

"More or less."

He turned thoughtful, plunging his hands deep into his pockets and looking out to sea. "There's groupers out there," he said. "Definitely. I can smell fish, and I guarantee you, friend, this place is *fishy*."

The next morning the sky was finally clear and the water blue and calm. I peeked out the door to make sure the fisherman was not still lying in wait for me, then hoisted my net bag of diving gear onto my shoulder and headed out to the parking lot. In the fierce morning sun I climbed the ridge to the Smithsonian station again, the underpowered little scooter nosing incrementally upward as I held the throttle all the way back. As I climbed, the sea expanded in all directions, shimmering and sharply defined against the parched, ragged edges of the island. I thought of John Glenn's space capsule, falling from the darkness of orbital space toward this luminous ocean, this smaller pool of infinity.

Matt and Jim, two members of the Smithsonian team, had a day off today, and I had arranged to go diving with them along the shallow reef crest at the edge of the windward lagoon. Matt was waiting for me, standing in front of the quonset hut along the conch-lined driveway, already wearing his weight belt. He had a big knife strapped to it, Mike Nelson style.

"Killer day," he said. He was in his early thirties, a botanist. He wore a black T-shirt with Nazi-style lettering advertising some sort of motorcycle charity event, with a picture of a Hell's Angel lookalike giving a waif a ride on his chopper.

When Jim arrived we carried our cumbersome gear down the steep boardwalk that ran along the bluff to the boathouse. Jim was a few years younger than Matt, quiet and studious. He wore a yellow bandanna on his head. Last year, he said, he had worked in the Everglades, collecting plant specimens for Biosphere II, the self-contained habitat in the Arizona desert that was then under construction.

We loaded our tanks into a small outboard Whaler and cast off, Jim in the stern piloting the boat and Matt and I standing in the bow, keeping a watch out for coral obstructions. Today the water looked not only clear but scoured, as if all the storm-riled particles of sand from the night before had merely ground the lens of the lagoon to a more brilliant tolerance. I saw stingrays moving lugubriously away from the boat, and now and then a small shark, propelling itself across the sand fields with a wriggling stroke.

Whenever I saw the blue tongues of coral formations intruding upon the white sand, I waved Jim around them, but as we got closer to the reef the sand channels narrowed and the coral compacted around them in an almost solid maze. Here and there jagged pieces of elkhorn coral broke the surface, looking from a distance like corroded steel, and the water swirled and sucked around them as if excited by an electric current.

I saw how easy it would be to run the boat up onto one of these coral heads and rip open its hull. The sea was booby-trapped. Just beneath the surface of the ocean was another surface as ragged as a cheese grater, against which a ship and its unprotected crew could be ground into fragments. The waters of the Turks and Caicos are filled with such menaces — low-lying reefs that lurk at the margins of the deep ocean passes like sand traps on a fairway. From my reading of Herbert Sadler's history, I remembered the names of some of the ships that had sunk or foundered or simply disappeared: the *Soleil Levant*, the *Invincible*, the *Eastern Light*, the *Nathan Hale*, the *Fortuna Pieper*, the *Zephyr*, the *Bella Dolores*, the *Anne*, the *Jeanne*, the *Miriam*.

The earliest such wreck is nameless. No one knows the identity of the Spanish vessel that went down in 1513 or thereabouts on Molasses Reef, a notorious ship trap a few miles southwest of Grand Turk. From circumstantial evidence — the size and shape of its wrought-iron ordnance, the length of the disintegrating timbers, the amount of ballast stones — archeologists have speculated that it was a small exploratory vessel like the naos and caravels used by Columbus. Such ships, the first reliable craft to venture across the Atlantic from Europe, were descended from the fishing boats and coastal traders that had been refined over generations by Iberian sailors. As the Spanish conquest of the New World intensified, larger and heavier ships — built for colonizing expeditions or for the transport of treasure back to Spain — began to replace the nimble vessels that had scouted the way.

The wreckage of several of these latter-day cargo ships has been discovered, most notably the three members of a Spanish treasure fleet that went down in a storm off Padre Island in 1554. But only three wrecks preceding that date have been found, and none has been positively identified. As a consequence, we know very little about the ships that bore those first European explorers to the Antilles and beyond. We know more about the Pharaonic barges that cruised up the Nile 2,500 years ago than we do about the *Niña,* the *Pinta,* or the *Santa Maria.*

The Florida treasure-hunting company that first located the Molasses Reef wreck in 1980 claimed that it was in fact the *Pinta,* whose fate is unrecorded after it returned to Spain along with the *Niña* at the end of Columbus's first voyage. (His flagship, the *Santa Maria,* broke apart on a reef off the coast of Hispaniola. Never found, it remains a holy grail of underwater archeology.) The treasure hunters' claim evaporated for lack of evidence, but several years of fieldwork and research by the Institute of Nautical Archeology, a nonprofit organization based at Texas A & M University, has indicated that the Molasses Reef wreck is the oldest sunken ship ever found in the New World.

I had missed the 1982 excavation in the waters of the Turks and Caicos, but years later I drove up from my home in Austin to one of those raw new exurban communities near the Dallas–Fort Worth airport, where the eight tons of artifacts that the archeologists had recovered — firearms, anchors, fasteners, ballast stones, nails, shot, and cheesy, waterlogged fragments of timber — were being stored pending delivery to the new national Museum of the Turks and Caicos. A two-story tract house on a residential street served as the headquarters and laboratory of Ships of Discovery, an organization founded by Donald Keith, the archeologist who had led the 1982 Molasses Reef excavation.

Ships of Discovery's purpose was passionately specific: to find the vessels that had been lost during the early years of New World exploration, ideally the very craft — the *Santa*

Maria or the *Gallega* — that Columbus himself had lost during the course of his four voyages to the Indies. Keith was not interested in merely excavating random shipwrecks; he wanted to catalogue an era: the climax of the age of sail.

There was no indication that the Molasses Reef wreck had any direct connection to Columbus. Months of research in Seville — in the Archivo General de Indias or in the Archivo Protocolos, a dreary old church whose high shelves were stacked with notary records from Imperial Spain bundled in ancient leather wrappers — had revealed no provenance for the ship. It was a mystery wreck. The artifacts themselves provided mostly vexing clues to its identity and its purpose. Historians had always assumed that exploratory vessels were lightly armed, but twenty pieces of artillery had been recovered from the Molasses Reef ship, along with five pairs of ankle shackles. Why so much ordnance? Perhaps, Keith and his colleagues speculated, it had been an outlaw slaving ship, ferrying captive Tainos from their home islands to their miserable destinies elsewhere in the Caribbean, and bristling with firepower in case it encountered another lawless vessel such as itself.

The Ships of Discovery headquarters had been rather eerily converted from a family home into an archeological laboratory. I wondered what the neighbors thought was going on next door, where bearded young men came out periodically to stir a tub of heated microcrystalline wax or to step into a prefab tool shed to check on various mysterious objects soaking in a series of electrolyte baths. The two-car garage was filled with crates and file cabinets and clear plastic sweater boxes containing bits and pieces of the wreck — flat-head nails, straps, bolts, drift pins, and cast-iron shot the size of tennis balls. The few remaining wooden fragments of the ship's hull were tightly stored in gauze and bubble wrap. I put on a pair of white cotton gloves in order to run my hand along the length of a wrought-iron swivel gun known as a *verso*. It was four feet long and prehistoric-looking; it might have been the charred femur of a mammoth. It had lain underwater for over four hundred and fifty years, but it was still in one solid piece, and when I touched its black surface I could feel the ripply hammerstrokes put there by the man who had forged it.

The *verso* had not looked like this when it was first discovered underwater. Over the centuries it would have grown a perfect camouflage coating of coral and been indistinguishable to an untrained eye from the reef itself. The popular image of an ancient shipwreck — a moody, moldering hull looming above the ocean floor — is a fantasy. When a wreck is as old as the one at Molasses Reef, chances are the wooden hull has long since rotted away, the anchors and artillery and chains and other prominent features are hidden by encrusting marine life, and nothing whatever is visible.

I had seen one such wreck at first hand, when I had gone with Donald Keith and his Ships of Discovery crew, along with a team of Mexican archeologists, to the site of a sunken ship near the resort city of Cancún. It was known as the Bahía Mujeres wreck, and Keith thought it was the same vintage as the ship that had sunk off Molasses Reef, which made it one of the oldest shipwreck sites yet discovered in the Western Hemisphere. But the identity of the Bahía Mujeres wreck, like that of the one on Molasses Reef, was unknown, and the site had been heavily disturbed since its discovery in 1958. Most of the major imperishable features of the ship — its cannons and anchors — had long since been hauled off.

The wreck lay in six to ten feet of water not far from the Cancún shoreline, which was crowded with gleaming white hotels, three-tiered sightseeing boats, and ski boats trailing parasailors and crowds of children on giant inflatable cylinders.

"You are coming with us today?" one of the Mexican archeologists asked as I pulled on my wetsuit. "Then you will learn a new dance!"

I saw what he meant as soon as I was underwater. *Marejada,* the surging sea. We were in ten feet of water, and the swells overhead were six or seven feet high; when I looked up from the bottom I saw the waves traveling overhead in peaks, and I felt their force when they broke up on the reef in a boiling underside of foam. It was impossible to stay in one place with the surge shaking me like a rag doll and a steady underwater current nudging me, like some sinister presence, toward the open ocean.

As for the Bahía Mujeres wreck itself, I saw nothing: just another coral garden, partitioned with string into grids, with a few shallow archeological trenches cutting through it. I peered hard at the coral as the watery ceiling slammed over my head, searching for some anomalous detail, some shape or angle that would not have appeared in nature, but the deception was total. An immense school of blue tang swam by, moving like a scaly wall, and a group of yellowtail snappers drifted in the surge with a kind of narcoleptic indifference, riding out the rhythmic ocean storms without moving a fin.

I lent a hand where I could, but the work was mysterious to me. My first job was to keep the slack loops of the measuring tapes from tangling in the coral while a triangulation team measured distances and angles in order to plot the site on a map. Then I helped Don Keith carry a plastic milk crate full of diving weights from one side of the site to the other. The weights were markers, to be placed wherever an artifact was found. Small Styrofoam buoys had been tied to them, and they sprouted from the milk crate like a bouquet of long-stemmed flowers. When we were through moving them, Keith set the box down and mimed the act of wiping sweat from his brow.

I made my way over to one of the trenches, where several divers were chipping their way through the rock with hammers and picks. They had tied their fins to their belts — you couldn't swim in this surge — and they wore a lot of weight to keep them hugging the bottom, but even so they had to stop working every few seconds and grab hold of something as the surge channeled through the trench like a flash flood. One of the archeologists was attempting to make a sketch on her Mylar slate. She was assisted by another diver, who held onto her weight belt and planted his feet against the gusting water. With the strange undersea weather howling around them, and all that rock to be laboriously chipped away, the archeologists made me think of prisoners toiling away in some penal colony on a distant planet.

When I finally saw the shipwreck, it was only because one of the archeologists pulled me into the trench and showed me a cross-section of the ballast mound. I would never have noticed it on my own, but now I perceived that beneath its coralline surface the rock had a cobbly texture. It was an accretion of smooth stones from some river far away in Spain.

This ballast mound — these rocks — was basically all that was left of the ship that sank here early in the sixteenth century. In those years, when ships were built by hand and eye rather than precision machines, ballast served a crucial navigational function. Since the hull of a ship was likely to be a bit asymmetrical, permanent ballast acted like the balancing weights on the rim of a car tire. The stones were large and were as much a part of the ship as the rigging or the sails. On top of them went the temporary ballast, smaller stones that could easily be tossed to and from the dock, depending on the size of the ship's cargo.

The way the ballast lies after a shipwreck tells a story. On the Bahía Mujeres wreck, for instance, the central ballast mound was suspiciously small; the stones were not all congregated in one place but dispersed over a wide area. This may mean that the ship did not sink at once, but struck the reef and foundered. The crew would have jettisoned the vessel's artillery and ballast stones in an attempt to lighten it and float it off the reef. But obviously they were not successful.

As I helped guide our little boat through the coral at Grand Turk, that ballast mound from the Bahía Mujeres wreck loomed in my mind. I wondered how many undiscovered wrecks lay at the bottom of this very reef, their planking eaten away, their iron parts festooned with coral, nothing to commemorate their destinies but a pile of egg-shaped stones.

We anchored next to a big coral head just inside the reef line, a place Matt and Jim referred to as Algal Ridge. The water was only ten or twelve feet deep here, and as soon as I slipped overboard I could feel it crashing and surging over the shallow coral bank. The movement of the water left me alternately hanging in a vacuum or driven forward in a churning rush, past the soft branching corals that shuddered and swayed like trees in a cyclone. I followed the surge the way I would follow the current of a river, letting it lift and propel me wherever it wanted me to go. There was no point in fighting it, and I liked the flotsam feel of being swept along the flanks of the massive coral heads. These growths of coral were high and fissured, looming above the sand flats like mesas rising from a desert floor. And like mesas they became larger and more complex as you approached them. What seemed from a distance to be a simple loaf-shaped coral rock only ten or twelve feet long was suddenly as big as a house, made up of intricate valleys and spurs and meandering caves.

When he reached the bottom, Jim pulled out a Ziploc bag and began collecting algae for the larger crabs. The red algae known as *Dasya* was everywhere, either moored to the rocks or torn free by the surge. It floated past us incessantly, a drift of fine pink feathers. Jim also plucked green *Halimeda* from the crevices, where the algal growths had anchored themselves by means of a fibrous knot called a holdfast. Algae do not have roots or stems and, though they often form stout paddles or blades, they do not have true leaves. They are the earth's most ancient plants, simple aggregates of simple cells with an ingrained hunger for sunlight. Through photosynthesis, they transform this absorbed light into the nutrients upon which all marine life is premised. Without them there would be no coral, no fish, no reef, no idling humans. Underwater, algae are everywhere, growing in mats or sprigs or in delicate tracery on the coral rock, but the average diver may, like me, tend not to notice them at all unless he is diving in a kelp bed in California, where the algae grows in soaring beanstalks 150 feet high, blocking out the sun and permitting only a dappled forest light to settle into the depths.

For a few minutes I conducted a dutiful inspection of the various algae species growing in the vicinity, but I was soon distracted by the more invigorating spectacle of the inverted waves building and crashing overhead. The reef crest is the front line, a zone of continuous violence and turbulence where the pummeling force of the water holds most species of coral at bay. This is the domain of elkhorn coral, which grows in an artful tangle of flat, swept-back branches that help to dispel the force of the oncoming waves and to present the elkhorn's polyps to the vital sunlight.

The elkhorn looked beautiful to me as I swam along below it. The branches stood unyielding in the agitated water, and the densely packed burrows of the individual polyps gave their surfaces a scarred and ancient look, like old bones that have been exposed to the sun and grown brittle and pockmarked. Indeed, the growths of elkhorn reminded me of a prehistoric fossil heap, where some herd of Ice Age creatures far larger than elks had died en masse, leaving behind only a twisted nest of giant antlers.

There were few fish around, but beneath the pale yellowish elkhorn canopy the other colors of the reef were vibrant — purple sea fans, fireworms with blood-red stinging needles, tiny cleaning shrimp with a candy-stripe pattern of red and white. Caverns were everywhere. I stuck my head into a crevice to have a look around, expecting an indentation in the coral of only a few feet, and saw instead that the whole bank was honeycombed with corridors and tunnels. It was irresistible, and I swam inside. Rooms and rooms opened ahead of me, the kind of ever-expanding interior space one encounters — full of rapturous expectation — in a dream. The floor was clean sand, with precise ripples that charted the direction of the water flow. Light came in skittery beams from holes in the ceiling. And every room I entered led to some deeper chamber, from which tunnels radiated in all directions, leading either to more rooms or to an exit marked by a pane of blue ocean light. The walls, scoured smooth by the relentless pulsing of the water, were festooned with algae and coiling sea whips and tubular sponges.

Through the holes in the ceiling I could see the spreading elkhorn branches, benevolent as the shady limbs of a great oak, and the water crashing over them. The surge caused a constant high-pitched whine, but otherwise there was silence. I saw Matt entering through an opening up ahead. He waved at me and disappeared into another room. I waited for the gathering power of the surge, and when it came I let it propel my body through the labyrinthine passages of the reef in one fluid rush. It was like surfing underwater.

I settled into one large round room and stayed awhile, watching the light beams shift across the sand floor. A school of margate, a species of grunt, swam tentatively past. Grunts are night-feeding fish with forked tails and down-slung eyes. During the day they while away the hours by closing ranks and slinking about in densely woven schools. To me, floating about in my chamber like an underwater lord, they looked morose and craven — and fascinated with *me,* of course. They seemed to be pondering what move I would make next. I suddenly remembered an odd scene from James Jones's Caribbean novel *Go to the Widow-Maker,* where the protagonist, finding himself alone in an underwater room like this, promptly pulls down his swim trunks and masturbates. I would not perform that act today for these prurient margate, but I saw the strange romance of it, the compulsion to release one's seed into this coral womb, to watch it float upward in ropy strands, one more hopeful clutch of waterborne protoplasm searching for its destiny.

The grunts passed on, but I stayed awhile longer, hanging in the center of the room in perfect trim. The surge was gentle here in this hidden space; its rhythms were predictable. I closed my eyes and felt it rocking me. When the needle on my gauge began to nose into the red I headed out, wandering in no great hurry through the coral labyrinth. At this depth I was not using much air, and at any rate I could see the boat through the portholes in the cavern ceiling. It danced on the surge, the anchor line alternately straining and relaxing, the waves sweeping under the hull with billowy momentum. Every once in a while as I swam through the sunlit caverns, I would catch sight of Jim or Matt popping into sight and then disappearing into another room. It reminded me of one of those old movie routines where people are forever missing each other as they come and go through the doors of a long hotel corridor.

Jim was waiting in the boat when I surfaced, and Matt came along soon after, spitting out his regulator to talk as he hauled himself over the gunwale. "Did you see that giant grouper?" he said. "Oh, man, I wanted to take out my knife and stab him. I wanted to kill that sucker with my knife."

Matt's face was lit by a giddy smile. Jim and I looked at him and laughed. The killer botanist. But we were all giddy, and I understood Matt's desire to take a trophy from this spot, to possess something from the reef that would be more durable than our already fading memories.

As I hauled up the anchor I looked down at the coral heads we had been swimming through. Now they appeared merely as subsurface hazards again, jagged and grasping forms that could snare our little craft the way their millions of constituent polyps could snare a passing bit of plankton. I imagined that long-ago scene on Molasses Reef, a moonless night perhaps, or a washed-out morning with the flat light obscuring the horizon, the sea calm with no foam marking the dangerous line of the reef. An outlaw caravel sailing in the lee of the Turks and Caicos to avoid the strong easterly winds that would otherwise threaten to blow it to Cuba; the crew watching for the deep-water passage leading to the open Atlantic; the blacksmith pouring lead into a shot mold next to a glowing brazier; the Tainos in shackles, sullen with the unimaginable weight of the tragedy that has befallen them. And then the clear coasting momentum of the ship brought to a wrenching stop so suddenly that the crew pitch forward against the deck, knowing as they fall that their vessel is gone and that the most they can hope for now is to somehow make their way to a nearby scrub island and spend the next few months or years trying to survive until they are delivered. They see the reef as the enemy, the invader, tearing into the protective shell of their ship the way a Carib arrow might tear into their bodies. They do not know, nor could they take comfort in knowing, that within these coral mounds there are beautiful passageways, beveled rooms full of dungeon light and waving sea fans.

10

Faceless

When I first came to Grand Turk, I had a problem with fish. I didn't like them much, didn't respond to them, didn't understand what they were supposed to be about. Fly fishermen are always chattering on about the nobility and mysterious cunning of trout, and the dedicated aquarium keepers I know seem to regard their fish tanks as an almost spiritual locus. But when I used to hunt for redfish and sea trout in the waters of Laguna Madre I regarded my prey with cold disinterest. Looking back, I suspect that this was only a boy's way of demystifying the creatures he has sworn to kill. Denying my fascination with them was a way of distancing myself from the guilt and emotional trauma of destroying them. I willfully embraced the popular notion that fish do not feel pain, and when I watched them expiring on the deck, their gills flapping and their mouths puckering in silent agony, I convinced myself that these were mere mechanical spasms, no more "felt" than the shuddering of a dying engine. I had heard somewhere that slipping an ice cube into a fish's mouth would kill it instantly, and I tried this over and over without success, determined to believe in this magic shut-off valve.

Like most people, I usually encountered fish when they were dead or dying — trussed up on ice in a fish market or hanging limply from a stringer off the side of a boat or pier. In their sameness, their helplessness, their mute acceptance, they struck me as pathetic. There was no recognizable warmth or passion in them — no *purpose*. And the exotic living specimens I saw displayed in the aquarium tanks of zoos and oceanariums were simply animate baubles, to be admired for their pretty colors but never considered as sentient beings. I was drawn to moist-eyed mammals or to forthrightly repellent reptiles. The only fish that interested me were the fish I had some overt reason to fear — sharks or electric eels or grotesque poisonous stonefish.

When I began diving in the ocean I should have seen at once how puny my imaginative responses to fish had been, but for years I squandered my opportunities to observe them in their rightful context. I watched them dart across the reef, full of their urgent business, and I learned most of their names, but my attention was arid and dutiful. My imagination refused to reach out to them.

This was my state of mind during my first few weeks of diving at Grand Turk. I would float at the promontory of Harmonium Point and list the passing fish on my slate as if I were recording the details of some exotic procession. Here was a queen angel, a rock beauty, a hogfish, an immature bluehead wrasse. Duly noted. But I kept straining my eyes past them, hoping to see some creature I recognized as belonging, in some unknown way, to my own kind. I wanted to locate a brute mammal — a manatee, a humpback whale — calling to me from across the void.

Then one day a school of creole wrasses swam by, following the lip of the wall. There were so many of them that they filtered out the sunlight, and the water grew darker as they passed. Wrasses — whose many species include the hogfish, the bluehead, the yellowhead, the clown wrasse, the puddingwife, and the slippery dick — make up one of the most common families of fish on the reef. They tend to have colorful and rather blunt bodies, with a frilly ridge of fin along their backs. Creole wrasses are larger than most of their wrasse cousins, and stockier. They are usually blue, with a black nosecone marking covering their forehead from their eyes to their upper lip.

Something about this school swimming by attracted me, and I joined them, rising from the promontory and coasting along in a flow of eight-inch fish. They swam in perfect unison and perfect order, moving forward with sweeping motions of their pectoral fins, each one with a focused and unquestioning will. I was as exhilarated as if I had flown up into the sky and joined a flock of migrating geese. Soaring along with these wrasses, I felt myself warming to the idea of what a fish might be. Every member of the school had the same dour and determined look, the same lifeless eyes, but for the first time this sameness struck me not as threatening but as beautiful. I saw that the fish were not so much individuals as particles, the spangly elements of a mysterious scattering layer that seemed to be constantly shifting in quest of one ultimate coherent shape. The school moved like a dense current, with small groups breaking off here and there to form swirling eddies that eventually found their way back to the main channel. Swimming along with the wrasses were smaller fish, blue and brown chromis, and on the flanks of the school carnivorous jacks and barracuda kept watch like malevolent shepherds, waiting for their chance to attack.

Beside me, inhabiting the same plane, swam a Spanish mackerel, its black dorsal fin tipped with white.

I dropped down deep into the center of the school and followed the wrasses as they trailed between twin pinnacles of coral. When they passed through the pinnacles they swam upward and then down again, moving across the reef in an undulating wave. Swimming with them, borne along on this tide of fish, I felt eerily happy. Unbidden, the theme from *Gone with the Wind* kept playing over and over in my head as I swooped through the reef formations with the wrasses. When the fish in front of me suddenly swerved in another direction, and the light hit their scales from a sharp new angle, it was as if they had been struck with a magic wand. They lit up in a sparkling explosion of blue.

The school moved forward as if it had a destination. Creole wrasses are plankton feeders that tend to patrol the midrange waters of the reef in the brightness of day, sinking down at night to burrow into the sand and fall asleep. I assumed they were feeding now, plucking jellyfish and copepods out of the water, but I could not get myself to focus on one fish long enough to be able to tell. And this is probably the primary reason why fish congregate in schools — to frustrate predators, to hopelessly splinter a potential attacker's visual cues. I remember once diving through a tunnel in Grand Cayman that was filled with a school of small, glistening fish known as silversides. They appeared to be not just a congregation of individual creatures but a solid mass, and as I stroked through the tunnel I could see nothing else, just a glistening silver curtain that kept parting endlessly in front of me. If I made a sudden gesture, if I tried to reach out and touch a specific fish, the school exploded in a blinding starburst in which my target fish was endlessly refracted.

Threatened with attack, a school of fish becomes something like an optical illusion, a mirage hovering continually out of reach. These creole wrasses were not as spectacular in their schooling behavior as the silversides I'd seen in that tunnel, but they clearly subscribed to the notion of safety in numbers. Here on the seaward margin of the reef, where the big predators were on the prowl, they could search for food in relative peace of mind, able to instantly congeal into a pointillistic whole and whoosh out of an attacker's way like a matador's cape.

No one knows exactly how schooling fish communicate danger with such precision and alacrity. Every fish in the school appears to receive the message at the same time, as if wired into some central power source. It is known that fish have rich sensory lives. They are able to hear, and they communicate with one another by all sorts of chirps, grunts, and grinding sounds. Their eyesight is good, and most species see in color. The lenses of their unblinking eyes are thick and bulbous to bend light waves traveling through the dense water and direct them onto the retina. Some fish have excellent binocular vision; others — those whose eyes face outward from the sides of the head — apparently perceive the world as two more or less distinct lobes of sight. (Creole wrasses belong in that category. Their eyes can move independently of each other — one eye locked on a floating morsel of food, the other scanning the opposite horizon for danger.) Fish have taste buds not only on their rigid tongues but often in the probing barbels that hang down from their mouths, and in some cases taste buds are scattered all over their bodies. Some species of catfish, for instance, can taste with their tails.

Fish, by and large, are prodigious smellers. As with humans and other vertebrates, the neurons that convey a fish's sense of smell report directly to the complex areas of the brain that govern emotion and behavior. Living in the water, a fish is steeped in dissolved smells. For many, the sense of smell is what provokes them the most, spurs them to "higher" responses like sexual desire or fear. In laboratory experiments, male gobies have gone into courtship displays when the water in which a breeding female had been swimming was poured into their tank.

Smell helps lead fish such as salmon back to the rivers and streams of their birth. Smell helps fish find food and perceive danger, and it probably is a factor in the impulse that gathers them into schools. But other sense organs are important in schooling as well, particularly the lateral line.

Generally, fish perceive the world in a way that we can understand. Like us, they hear, see, taste, smell, and have at least a dull sense of touch. And we can speculate, at least, on a fish's mental awareness — a brain full of sudden decrees and impulses that do not linger long enough to be considered thoughts. But we are unequipped to grasp the sensations that come to them through the lateral line. This is a sensory canal that runs along each of the fish's flanks. It is studded with sensitive motion detectors, which constantly monitor low-frequency vibrations. The lateral line allows fish to "hear" changes in water pressure, to gather information about other creatures by the way their movements disturb the water. Because of the lateral line, fish in a school are able to swim at an unvarying distance from one another, and to view the ocean space between them and their neighbors as a tingling web of perception.

The school of creole wrasses, the more I swam with it, struck me as a lovely work of architecture, a solid thing created by the dynamic tension of swarming molecules. But the individual fish still filled me with a vague anxiety, and I realized that my lifelong disenchantment with these creatures might have been based on the fact that they have no faces. Fish do have certain fixed looks that can strike us as familiar or even endearing. A French angelfish has a soft and expressive visage. An African pompano, I have observed more than once, bears a direct resemblance to J. Edgar Hoover. And if you look at the pale undersides of certain species of rays and skates you will see what looks unnervingly like the face of a human baby. But these are masks. Fish have no facial muscles, no way to vary their frozen expressions. They can look permanently startled, quizzical, wrathful, serene, but their countenances are accidents of evolution, of ages of hydrodynamic trial and error, no more reflective of the fish's interior state than its fins or scales. That was why, as a boy, I was able to convince myself that they did not feel pain — because they did not seem to *care*. But this lack of caring, this pokerfaced nonchalance, frightened me more than it comforted me. I think I worried that it was contagious. I had a horror of being faceless myself, of being numb and indistinguishable and marching with hypnotized unconcern into some dreadful trap. Fish to me were the brainwashed legions, the living dead.

Now, watching them underwater, I began to understand that fish were as fervent about their lives as I was about mine. I peeled off from the school of creole wrasses and saw them disappear into the enveloping distance like a haunted caravan. As I headed back toward the boat, I was interrupted by the attack of a damselfish about six inches long. The damselfish didn't bite, but would have if I had persisted in trespassing on its tiny patch of algae, which it was cultivating on a tableland of dead star coral. More than likely the damselfish had helped kill the coral, nibbling away at the polyps until they were gone, like a farmer ridding his pasture of tree stumps. Many species of damselfish are precisely that: farmers. They grow algae and they eat it. Their fields are usually minute tufts of green, which the fish patrols frenetically. At the approach of any stranger, no matter how large, the damselfish attacks. Damselfish are everywhere, and if you swim low over the reef, you are constantly rousing them to action. They make heedless lightning charges, and they will not be repulsed. I backed off just outside of this damselfish's boundaries and studied its fretful possessiveness. When a big parrotfish swam by, the damsel took after it with the ferocity of a water buffalo, and the parrotfish veered off without hesitation. Why it would be intimidated by such a slight adversary I had no idea, but the confrontation had the air of ritual, of a dance whose movements were so deeply encoded in both fish that it transcended my understanding of natural logic.

More parrotfish streamed by, potential claim jumpers that gave the embattled farmer no rest. It was a bright afternoon, and fish were everywhere in schools or loose federations. I counted the species within a three-yard radius of where I was swimming: trumpetfish, black durgons, creole wrasses, striped parrotfish, stoplight parrotfish, blue chromis, bar jacks, yellowtail snappers, goatfish, banded butterflyfish, bluehead wrasses, fairy basslets, assorted gobies and blennies, trunkfish, cowfish . . . I gave up. It was suddenly astonishing to me how visible it all was. What if the terrestrial world were like this? What would it be like to take a stroll in the forest and find, within an arm's-length radius, a similarly teeming display of woodland creatures: beavers, robins, woodpeckers, deer, squirrels, chipmunks, rabbits, foxes . . . ?

A coral reef may be the most industrious, pulsating, *driven* environment on earth. Something was always going on here, most of the time right in front of my face, yet the action was so cryptic and so furious that I could not track it, and much of what I happened to see I could not comprehend. But the reef in its vibrancy appeared to have nothing to hide. Many of the fish I routinely encountered — like the various parrotfish and several iridescent species of gobies — were turned out in what Konrad Lorenz called *plakatfarben*, "poster colors." And even fish that I thought of at first as drab had, on close inspection, a radiant undercoat that shone through their scaly hides like some deeply woven design rising from the fabric of a carpet.

I followed a black durgon for a while, attracted by the breezy, rippling motions of the dorsal and anal fins that framed its flat, high-beamed body. At first glance a black durgon is not a colorful fish — it is black. But now I noticed the color lines highlighting the base of each fin, two silky blue stripes as rich as the lining of a tuxedo. And the more I looked at the durgon's black background, the more I saw a shimmering play of colors — a spectral blush beneath the skin, as unexpected as the rainbows that sometimes form in roadside oil slicks.

On the reef you cannot get away from color, from the infinitely varied play of water and light upon shining surfaces. Sometimes the display is lurid, sometimes so subtle you see it only as a wavering illusion. For fish, color often functions as camouflage. Like the octopus, many species of fish can alter the pigment cells in their skin to match their background. Cruising above the sand flats, I learned to spot motionless peacock flounders by the oval patterns they made in the sand; the fish themselves were almost invisible, their tan skin accessorized with faint blue rings that imparted a suggestion of flickering surface light. Flounders, like octopuses, are among the supreme masters of camouflage. (In one laboratory experiment, a flounder managed to produce a crude imitation of the checkerboard background on which it had been placed.) Just as impressive in their covert survival strategies are all the warty bottom dwellers like frogfish, toadfish, stonefish, and scorpionfish. These tend to be creatures with cavernous mouths, sometimes with fleshy lures attached. They look like slimy hunks of coral or encrusted rocks. When a prey fish swims close enough, drawn by the wormlike lure, it is inhaled with breathtaking speed, sucked into the bony, hollow darkness of its ambusher's mouth.

The most notorious such animal is the Australian stonefish, which lingers in shallow waters, indistinguishable from the coral rubble. It has thirteen venomous spines along its back to fend off, among other things, the human foot. I grew up believing that if you so much as touched a stonefish you would experience three to five minutes of unimaginable pain followed by a certain and merciful death. This is nine-tenths folklore. It's true that people occasionally die from stonefish encounters, but while the venom is consistently agonizing, it is not often fatal.

Here in the Caribbean, the local variant of the stonefish is the scorpionfish. The neurotoxic venom of a scorpionfish is not as powerful as that of its Australian counterpart — in a laboratory experiment, the venom of eight scorpionfish was required to deliver a fatal dose to a guinea pig — but it is definitely to be avoided. On the reef I became adept at spotting scorpionfish after Mitch showed me a few. When he first pointed one out I had no idea what he was calling to my attention. He kept pointing at a nondescript piece of spongy-looking coral, reddish in color and covered with all the usual growths and protuberances. It took me a long time to realize it was alive. Finally I saw one of its eyes, a blank, glassy disk. The fish was wedged at an angle against the rock, supported on its stubby pectoral fins. Mitch took off his snorkel and poked it, and the scorpionfish threw off a cloud of silt, swam a foot or so away, and settled down again, unconcerned and undetectable.

Scorpionfish and flounders use color to conceal themselves on the bottom. Many other creatures use it to tone down their visibility in open water. Dolphins, rays, sharks, sea turtles, jacks, tuna — all are countershaded, dark on top and pale below. Seen from above, their blue or brown dorsal surfaces blend in with the ocean depths. When viewed from below, their undersides are hard to distinguish from the white light of the sun pouring through the water.

For the bright reef fish in their *plakatfarben,* however, color is a way of standing out. For French angelfish, with their sparkling yellow spots, or for golden hamlets or heliotrope basslets, camouflage is not a central issue. Unlike the naturally sluggish scorpionfish, they do not survive by being inconspicuous but by being fast and maneuverable enough to dart into a coral hole at the approach of danger. And in any case, the fish that prey on them in the muffled crepuscular light of early morning and late afternoon tend to be color-blind.

The brilliant colors seem to be for the benefit of other members of the species, a way to flash sexual enticements or territorial warnings across the gloomy ocean distances. Many of the brightest-colored species, like the parrotfish, are programmed to change sex from time to time, and each new phase is broadcast by a variation in color that alerts the rest of the breeding population to the individual's new gender.

I passed over a cluster of tube sponges swarming with tiny blue gobies. To my eye — always restlessly in search of leviathans — these fish were usually about as interesting as gnats. But today I was arrested by their shining colors, and I sank down onto the sand and watched them for a moment, trying hopelessly to look through their eyes. What did they see, what were they watching for? I tried to imagine the intensity of those color messages and of all those other mysterious stimuli — sound, smell, lateral line excitations — that formed the known universe of a goby's perceptions. Would a zooming yellow stripe, glimpsed across the chasm of the sponge opening, provoke a goby to an intolerable sexual pitch? Would the head-on approach of a territorial rival, its ventral fins whirring, its wide mouth moving up and down in percussive gulps, send a charge through its body that I would recognize as fear? I wondered, too, how gobies regard the ocean itself. They are dense, low-slung fish, lacking the swim bladders that allow other fish to adjust their specific gravities and move up and down at will through the varying density of the water column. To a goby, perched on a sponge or a piece of coral or a bed of sand, its face and eyes permanently canted upward, the ocean overhead must appear as an unreachable vault, the way the heavens do to us.

I looked down at these darting fish, two or three inches long. My eyes crossed as I tried to keep track of their comings and goings, and my imagination wore itself out trying to conjure up some authentic sense of the tenor of their existence. I knew that many gobies are monogamous, that the male and female excavate a nesting burrow together in the sand, and that in some species the male slips into the burrow to stay with the eggs while his mate closes off the entrance with sand, releasing him a few days later to swim around a bit before sealing him up again. But what does this mean? How does it feel? Could the reflexive pair bonding of gobies loom as dramatic to them, as essential, as human love does to us? I didn't know. I felt sheepish in even speculating about it. Gobies in love! And yet, *why not?* We dismiss animals as instinctive beings, and we think of instinct itself as an unfeeling mechanical urge, an electrical command fired from a witless brain. But we only know what we observe; we see only the motions. The acts of fish may be in fact nothing more than instinctive impulses, but we can never know the specific timbre of those impulses or the way a fish feels when it is impelled by them.

I became attached to a certain blenny. It lived on the coral precipice of Harmonium Point, forty feet below the surface. I first noticed it when I was idling through a deep sand channel with my faceplate inches away from the silty coral rock. Only the blenny's head was visible. It peered out from its burrow, a horizontal hole in the coral which the fish fit as perfectly as if it had been machined into place. When I looked it up later in a field guide, I saw that it was a secretary blenny. I guessed its length at about an inch and a half, though only its head — with its down-turned mouth and bristly appendages known as cirri — was visible to me. Two yellow rings neatly encircled its eyes like the war paint on a Comanche pony. Inside its burrow the blenny waggled its head from side to side, joyless and rhythmic.

Blennies are closely related to gobies. They are, on the whole, a little frillier, a little stouter. Like the gobies, they have no swim bladders, so they tend to stay on the bottom. They swim the way a seal walks, awkwardly, with their tails lower than their heads. But some blennies are able to travel across dry land, hopping like frogs from one tide pool to another. Blennies are generally carnivorous, and a few species are aggressive, but when I put my finger in front of this one it did not snap; it merely withdrew into its burrow.

The secretary blenny merited only two lines in my field guide and was not mentioned at all in the other books I consulted, so its life ways remained a mystery to me. But time after time, day and night, I dove down to Harmonium Point to check on this particular fish. It was always there, lodged in its burrow, moving its head from side to side. I was drawn to it because of its deadpan expression, its reliable location, its mysterious *lack* of behavior. As far as I could detect, it never did anything. Its world was limited pretty much to this minute hole in the coral rock, in one of the thousands of nameless canyons and sand channels cutting through the Grand Turk wall.

That little blenny became a touchstone. Once I dreamed about it. In the dream, I swam down as usual to the fish's burrow but found it empty. The blenny was gone. Only its yellow eye circles remained, vivid against the darkness of the burrow, sweeping back and forth with a ghostly vigilance. But even when I was wide awake the blenny struck me as vaguely unreal, stranger than all the other strange things on the reef. Its flush position in its burrow and its woeful expression reminded me of some other imprisoned face that, for a long time, I could not call to mind. Then, after weeks of perplexed pondering, it dawned on me: Señor Wences! Señor Wences was a puppet that used to appear, decades ago, on various TV variety shows. He was nothing more than a surly face trapped inside a box, whose lid would open now and then to allow him to make some acerbic, rapid-fire observation.

When I remembered Señor Wences, I felt a momentary triumph: I had the blenny pegged at last. I had managed to assign it a role, an identity. In my mind it was already a character to put beside the damselfish's embattled farmer, or the barracuda's sinister outrider. After a while I found that without consciously meaning to do so, I had peopled the reef with my own creations. One by one, I had given each fish a personality, and the stronger that illusory personality was, the more the fish interested me. When a fish did not remind me of anything or anyone, when it did not express itself in a way that captured my attention, I ignored it.

This was, of course, rank anthropomorphism, but I could not stop it, any more than I could get the theme from *Gone with the Wind* to quit playing inside my head. Though categorizing creatures in this way is a simple human tendency, harmless and ineradicable, I became impatient with myself for trying to reduce all this alien grandeur to something comprehendible and comfortable. I longed for my pedestrian imagination to fall away, because finally it was what barred me from truly perceiving the reef.

11

Lobster Rock

Mitch piloted the boat toward the beachside mooring in front of his house. It was midafternoon, and the sea was almost flat for once, no winds to rile the surface or stir up the sediment below. Even though we had just finished diving for the day, the clear, smooth water taunted me. In the last few weeks, I had developed an unreasonable concern that when I was not underwater I was wasting my life.

We idled to a stop in front of the sea wall. There was some movement to the water here, and the boat rocked pleasantly as the shore-bound wavelets slid beneath the hull. As always, we began pitching the empty yellow tanks overboard. Drained of their heavy compressed air, they floated like buoys and drifted onto the sand. I climbed up to the top of the sea wall, Mitch hoisted the tanks to me, and I carried them across the street to the compressor shed in back of his house. Then I gathered up my own gear and headed back to the Island Reef on my scooter. Trembling with hunger, I stopped off at the L & L Bakery and bought a loaf of their homemade bread and three doughnuts to eat on the way home.

In my room I made a sandwich and ate it while pacing back and forth, then I grabbed my snorkeling gear and went down to the grassy beach. Melvin, the hotel handyman, was vacuuming the pool as I walked by. He said if I brought back a conch he'd clean it for me, and I could have it for dinner that night.

Down at the beach I waded into waist-deep water, put on my fins, and stroked through the turtle grass beds. I swam with my legs together in a high-spirited dolphin kick, my arms flush against my body. I kept my fingers together and curled forward, enjoying the way the flowing water parted around them and eddied inside my cupped hands, tickling the palms. As I swam along, my heavy fins slapped the surface, creating a frothy turbulence that attracted the attention of a half-dozen bar jacks, which followed in my wake with confused excitement.

Tiny white flowers nestled in the turtle grass, along with spotted snails the size of my thumbnail, which gathered in clumps on the chalice-shaped algae stalks. Beyond the tangly turtle grass beds spread neat rows of manatee grass, the blades stiff and precisely arranged in rows, as if they'd been planted by hand. Up ahead I spotted a big diamond-shaped depression in the sand and guessed that a stingray had been rousted out of its resting place by the commotion of my approach. I followed its tracks — the straight groove where the tail had dragged, bracketed by scalloped patterns in the sand where the wings had touched down — and found the ray fifty yards ahead. It was a big one, with a three-foot wingspan. The tip of its tail was gone but the stinging barb at its base was still intact. The ray's skin was as dark as steel. The creature rippled forward along the bottom, stirring up the sand with its wings as it fed. At its wingtip a bar jack hovered opportunistically, waiting to grab one of the crabs or mollusks the ray might scare up. The jack annoyed the ray, and it moved off, a low-altitude hop culminating in a skidding landing ten feet away. This time it did not try to feed, just nuzzled down into the bottom and covered its steely gray back with a dusting of fine sand.

On the island of Grand Cayman, far away from here on the other side of Cuba, there is a famous diving spot known as Sting Ray City. It is without doubt one of the oddest tourist destinations in the world, a place where you sink into ten feet of water bearing chunks of squid and wait for a flotilla of stingrays to swim up and assault you.

I had been there once, and looking down at this ray I remembered the creepy, urgent press of those creatures. When the dive boat I was on pulled up at the reef line of Sting Ray City, I had looked down and seen the rays cruising expectantly below, their diamond-shaped shadows rippling over the sand bottom, their bodies perfectly still except for the fluted propulsive motions at the edges of their wings.

"Remember," the divemaster said brightly, "these guys are squid hogs. They'll do anything to get that squid out of your hands."

He reviewed the various coercive techniques the stingrays had developed to wrest their favorite food away from us. He said we would be in no danger from the poisonous barbs at the base of their tails — these are primarily a defense mechanism that the rays use when they lie hidden in the sand — but in a feeding frenzy such as we were about to create, they might very well ram us in the head to dislodge our masks, pound us with their beating wings, or suck our fingers into the bony vacuum tunnels of their mouths.

Thus briefed, we slipped over the stern of the boat with

plastic bags of chopped squid concealed beneath our buoyancy vests. I felt strange and a bit apprehensive, but going to Grand Cayman and not letting a dozen stingrays slither over your body was like going to London and not seeing the changing of the guard.

Once off the boat, we drifted down to the bottom and landed near a vast population of garden eels swaying in their burrows like meadow grass. The stingrays came soaring in at once, arriving with a consort of yellowtail snappers. The rays were larger than I had expected, with wingspans of three or even four feet, and their undersides were a gleaming white. They circled us for a moment, elegant and threatening, all of them moving along on the same plane. I could see their expressionless yellow eyes and the pulsing spiracles that pumped water over their gills.

The rays were not polite for long. When I produced a piece of squid, they began jostling and prodding and smothering me with their wings. One of them headed right toward my face mask and flared up at the last second, revealing a double row of ventral gills and that strange toothless baby smile. I tossed it a piece of squid, but a yellowtail darted in and stole it, and the rays began crowding me even more, as if in resentment. I could feel them sucking on my head with their mouths, and the silky smooth undersides of their flapping wings began to feel obscene and menacing. It was like being mugged by extraterrestrials.

I thought I would never be able to look at a ray again without revulsion, but those quirky nightmare sensations vanished soon enough. Instead I remembered the unearthly majesty of the rays as they swam toward me, the way they cut through the element of water as if it had no density at all and was as thin as air.

Like sharks, rays are cartilaginous fish, meaning that their internal skeletons are made of cartilage rather than bone. ("Take a shark's skull," writes Brian Curtis in his ever-amiable classic *The Life Story of the Fish: His Manners and Morals,* "and you can whittle it like a piece of wood, for it is just one big block of cartilage.") Many bottom-hugging species of sharks have flattened sides and skirtlike pectoral fins, and in the skates and rays these features have evolved into what are, for all practical purposes, wings. Rays — like sharks, blennies, and gobies — have no swim bladders. They move up and down in the water column through muscle power and body design, and the wings give them lift and propulsion.

Here in Grand Turk I had seen many stingrays and eagle rays, but I had arrived too late in the year to see the giant mantas that cruise through these islands in the summer, when the water is at its warmest and blooming with plankton. It seemed I was always just missing manta rays. They are fairly common, particularly in the Gulf of Mexico, but I had never encountered one underwater. Once, riding in a boat about a mile offshore from Padre Island, I happened to look over the side and saw what I thought was an oil slick, a smooth black stain fifteen feet across that traveled along beside the boat for only a moment before subsiding into the blue water. I searched the horizon for an hour and never saw it again, but I knew it was a manta. Nothing else in the ocean was so broad, so glisteningly black, and so silent.

Manta rays can weigh two tons and measure over twenty feet in width. Their flexible wingtips taper down to sharp points, and sometimes when the mantas swim near the surface they raise these wingtips above the water, so an observer is likely to mistake them for two sharks swimming with unusual synchronicity. At other times they may leap out of the water entirely, a gesture that is thought not to reflect exuberance so much as an attempt to knock the parasites off their bodies.

Their mouths are cavernous, guarded by two waving, protuberant fins that from certain angles resemble horns and are responsible for the manta's common name of devilfish. Mantas do appear deadly, even satanic; with their vast open maws, and their black spreading wings, they could easily be mistaken as reapers of lost souls. But they are utterly harmless. Diving magazines are filled with photographs of divers swimming beside them or hitching rides on their backs, sometimes even using the suckerfish that habitually attach themselves to the mantas' flesh as convenient hand grips. Because they are so huge and graceful, and so blankly tolerant of human attention, mantas often move people who have swum with them into believing they have had a religious encounter. How could a thing so darkly beautiful be anything other than a kindly, watchful spirit? In Peter Benchley's novel *The Girl of the Sea of Cortez,* the title character even befriends a manta ray, which saves her life and carries her down into a secluded canyon, an Eden lying deep in the ocean.

It is an old dream, right out of *The Water Babies:* to travel on the back of a benevolent sea beast down to some secret underwater garden. No wonder divers like me are always searching out and swimming up to the largest ocean creatures — whales and dolphins, manta rays and sea turtles and giant plankton-eating sharks. No wonder there is a booming business in "dolphin encounter" facilities, where people are only too happy to pay forty dollars for the privilege of entering a pool with captive dolphins. Lodged deep in the human imagination is an overwhelming desire to slip onto such a creature's back and be borne away into the sea.

And what better escort than the brooding winged form of a manta ray?

The sand in which the stingray had buried itself was fine-grained and bright. I could tell by the height and direction of the ripples that the wind had picked up and was blowing from the northeast. Cruising along the bottom, I turned over and watched the swells pass overhead. Watching the undersides of the waves, I felt as if I were inside some vast billowy tent. Here and there in the sand I found foot-long conch shells, their big fluted openings facing downward, but when I turned them over they were usually empty, a rectangular notch near the top indicating where a human predator had punched through the shell with a chisel or a knife to sever the adductor muscle that holds the conch inside. Several of the shells were home to small fish, and one had been taken over by an immense hermit crab. The crab's body was out of sight, tucked away in the spiraling chambers of the shell, but its meaty, brick-red claws protruded dangerously, like the shredding talons of some half-revealed monster in a cheesy science fiction movie.

I kept an eye out for a living conch as I swam along. I was already hungry again, and the thought of a big batter-fried conch steak was becoming a serious mental distraction. A four-foot barracuda came cruising along beside me, its teeth exposed in a wolfish grin and its galvanized skin shining. When it moved in closer, I thought for a moment it might be considering an attack on my tank valve, since barracuda are said to be attracted to objects as flashy and silvery as themselves. Back in the old days, divers were cautioned not to wear jewelry underwater, since a barracuda might take a notion to grab a wedding ring and a finger along with it. This has happened rarely, if ever, but barracuda are so ferocious-looking that they were thought by some early divers to be even more dangerous than sharks.

"The barracudas patrolled . . . tirelessly," writes the pioneer diver Hans Hass in one of his vivid books, "back and forth, coming a little nearer to me each time they turned. It was a regular siege. But they had not yet made up their minds to give the signal for the assault. . . . I . . . had the feeling that I was not confronted by individual creatures but by a sort of superego, with a steadily mounting common will. I could see no escape and began to lose my composure. I started jabbing viciously, in all directions, at one fish after another, with the blunt end of the iron shaft."

Hass's panic is understandable, considering that he was one of the first people to dare encounter barracuda in their natural habitat, but today even the most casual diver or snorkeler would read that passage with a chortle. Today we realize that the truly disturbing thing about barracuda is not that they are "besieging" us but that they are ignoring us. Though some momentary gleam or glimmer of our equipment may attract them, we cannot hold their interest for long. When, after a few moments, this barracuda swam away from me, I did not feel relieved, just dismissed. Once again I had the sensation of being nothing but a tourist, walking down some immaculate boutique-lined street, while the real city remained out of my sight, humming and throbbing with its secret life in the back alleys and bazaars.

This afternoon I had a premonition that I might veer off the tour a bit and see things I was not meant to see. Here, for instance, around a porous coral rock, was a bizarre tableau. A half-dozen queen triggerfish hung nervously around a deep hole in the coral. They seemed unusually touchy, and whenever another triggerfish violated their space they would make a run at it and chase it off. Other fish were about as well, mostly rock hinds and groupers, either stationed motionlessly at the base of the rock or puttering slowly around it. What was going on here? These fish were all drawn to the rock as if magnetized. Diving on the reef, I had grown used to the aimlessness of the creatures I encountered, sweeping across the terrain with no apparent destination. Now here was a simple outcropping of rock that seemed mysteriously to matter to them and to rivet their attention. The scene had an air of expectation and suspense. At the surface I hyperventilated through my snorkel, preparing for the lengthiest underwater foray I could manage, then stroked down fifteen feet to the bottom. The triggerfish scattered as I dove through their ranks, then immediately regrouped. I looked into the deep hole in the rock that seemed to attract them, and saw that it was crowded with stationary fish. There were several queen triggers, a spotted moray eel, a squirrelfish, a hind, and a big grouper, all of them perfectly still except for the periodic wave of a fin or a shudder along the flanks. The grouper held center stage. It was wedged into the rock, its spines extended. It seemed to be lounging, indifferent to all the strange tumult and fascination of the fish surrounding the rock.

With my slate, I touched the moray on the snout. It didn't react. It could have been comatose. None of the other fish inside the hole moved either. Their stillness had a sensual charge to it, a kind of languor, as if they were in the aftermath of some sort of ecstatic frenzy so strong it had short-circuited their physical responses.

In reality, I soon realized, I had come across a cleaning station. The motionless fish inside the hole were being groomed by wrasses and gobies and small cleaning shrimp, creatures whose diet consisted of exactly those ectoparasites the fish craved to rid themselves of. Cleaning stations are a common sight on the reef to those divers who pay close enough attention to find them. But most of the stations I had seen during my time in Grand Turk were on a much smaller scale — a lone client like a Nassau grouper hovering with its mouth open while a wriggling fish or two flitted in and out. This one was a superstation. The cleaning bay was full and customers were backed up, waiting for their turn. I dove down again and again, peering into the hole in the rock as long as my breath lasted, fascinated by the absolute repose of the big fish and the ceaseless industry of the cleaners. Here, all the normal business of the reef was on hold. The small fish that might otherwise be the prey of these groupers and eels were swimming around inside their open mouths, plucking parasites and dead tissue off their tongues.

Cautiously, I put my hand inside the hole. I had heard that cleaners could sometimes be induced to practice their trade on divers as well as fish. But the little neon gobies that I saw swirling around inside the hole ignored my hand. They were busy working over the flared dorsal spines of a squirrelfish, and I did not have the breath to wait for my turn.

So I swam on across the flats, and I came to a coral head about twenty feet long and six feet wide. I took a breath and stroked down to it, swimming along its length in the hope of spotting another cleaning station. No luck, but I saw a flamingo tongue, a gorgeous carnivorous snail, rasping its way up the surface of a sea fan, eating the lavender coral tissues as it went and leaving a black track behind. The flamingo tongue was a florid peach color overlaid with a reticulated patchwork outlined in black. The color belonged not to the shell itself, but to the living snail, which covered the shell with a thin mantle of flesh. When I touched this mantle gently with my finger, the snail retracted it, producing what appeared to be an instantaneous chromatic washout, the sumptuous colors and patterns evaporating in an eyeblink, leaving in their place a cold white shell.

Below the sea fan a wiry, waving thing protruded from a hole. I took it to be a lobster's antenna, and on closer inspection found that it was. Looking into the hole, I saw four lobsters facing outward and at least a half-dozen others bundled up behind them. A few feet away was another hole, and near it still another. The coral head was a giant swiss cheese; all the holes connected, and every hole was crowded with lobsters. I began to count them, rising every few moments to take another breath. Every good-sized hole had at least forty lobsters in it, and there was no telling how many were cowering back in the unseen corridors leading from one hole to another. I stopped counting at 800: eight hundred lobsters gathered together inside a coral head no bigger than my room back at the Island Reef.

Caribbean lobsters have no claws with which to defend themselves, but they are alert, agile, and otherwise well armed, with a spiny carapace and two sharp horns projecting forward over each eye. They have been observed to participate in mass migrations across the ocean floor, walking single file for thirty miles or more, using their thick feelers to keep in tactile contact with the lobsters in front of them.

When I poked my head into one of the holes, the crowd of lobsters inside shuffled back warily, their feelers wagging. Their eyes, on the ends of stalks, looked like tiny berries, and I could see the fixed horns overhead. Though it is illegal to take a lobster while scuba diving or with the use of a spear gun, I was within my rights to grab one with my bare hands while snorkeling. But the only place to grab a lobster with any reasonable chance of success is its unprotected tail, and these lobsters faced prudently outward, presenting me with a tangle of bristling antennae and fixed armor.

I had never tried to catch a lobster and had no idea how to go about it, but I tried anyway, reaching and groping with my hand. The lobsters were so quick they seemed to magically evade me, shifting position so fluidly I could not even touch them. Out of breath, I rocketed up to the surface. I had no breath left to blow the water out of my snorkel, so I just spit out the mouthpiece and gulped air.

I wondered why I was so excited. I had never been an underwater hunter-gatherer, had never even shot a spear gun, and had in fact been revolted years earlier when I had watched a fellow diver spear a red snapper on a sunken oil rig off the Texas coast. The fish had been swimming placidly only a foot or two in front of him, and shooting it had seemed a monumental act of discourtesy. The snapper thrashed in disbelief at the end of the spear. Its blood curled out of its body like smoke — green smoke, because the light rays that produce the color red are absorbed within the first few feet of seawater. The outrage of the fish and the otherworldly color of its blood unsettled me. This was a hypocritical attitude, I knew: there was no moral distinction to be made between spearing a fish and hooking one with a rod and reel — or, for that matter, buying a box of frozen fish sticks at the supermarket. But the way these snappers had accepted our presence, witlessly swimming up to us, made me feel vaguely ashamed when my companion turned on them.

But now, having found this mother lode of lobsters, I began thinking of myself as a predator and calculating how I could find this place again. As I swam above the lobster rock, I noted that I was about a half mile straight out from the pool at the Island Reef. I would not tell anyone about this place, but I might come back from time to time, equipped with a noose to slip over the tail of a legal-sized lobster.

I made one more quick underwater survey of the rock and then headed back toward land, swimming with my mask half in and half out of the water, so that the distant shoreline danced on a chasm of blue. Passing over the turtle grass beds this time, I noticed that living conchs were everywhere, lurching across the bottom with elegant slowness on the powerful tonguelike muscles that protrude from their shells. This muscle is known as the conch's "foot," and the creature uses it to hitch awkwardly forward several inches at a time.

I remembered Melvin's promise to clean a conch for me, and I swam down to grab one, a considerably easier chore than capturing a lobster. The conch I chose was large, with a pronounced flare to its shell. The mollusk's eyes peered out from beneath the shell, two black stalks with yellow circles at the end that appeared to have been daubed on with paint. The eyes looked dead, but even so I avoided looking at them as I picked the conch up and carried it to its doom. It must have weighed five or six pounds, and I grew weary carrying it back to shore, dragging it along like an anchor. When I walked out of the water with it I saw that a single turtle grass flower was rooted to the shell.

When I showed Melvin the conch he gazed at it admiringly. He said there were three methods of getting the creature out. All of them involved breaking the mollusk's death grip on its innermost whorl of shell. We could place the conch in the Island Reef's freezer overnight, and in the morning the frozen flesh could be easily pulled out. Or we could drill a hole in the operculum, the horny growth attached to the mollusk that acted as a kind of door when the conch pulled back into its shell. If we then put a string through the hole in the operculum and hung the conch in midair from a rafter, sooner or later it would grow too tired to hold on, and the shell would drop to the ground, leaving the formless flesh of the mollusk dangling from its string. The most direct approach was simply to punch a hole through the shell near the crown and cut the muscle the animal used to hold itself in. This method marred the shell, but it spared me the thought of that hideous, pitiful thing hanging from a string or slowly turning into a gelid, icy mass in the freezer.

So as I watched, Melvin knocked a hole near the topmost spire of the shell with the base of a knife, then inserted the blade of a smaller knife and wiggled it around inside. He grabbed hold of the operculum, which he called the horn, and tried to pull the mollusk out, but it wouldn't come.

"He's a strong boy," Melvin said. He punched another hole in a different spot, and this time the creature came out in one fluid motion, a yellowish, veiny mass that reminded me of a fat-clogged heart. The two eyestalks dangled, the eyes themselves as implacable as ever. Between the eye-stalks hung another black appendage, half their length and stouter.

"What's that?" I asked.

"He's a stud," Melvin said. "That's his wubby."

Melvin cut into the coarse, tough skin of the mollusk and began peeling it away from the powerful foot muscle.

"I love conch shells," he said. "I got a shell on my coffee table at home. I cut a hole in one end of it and put me one of those slim light bulbs inside, then ran the cord through the hole. I turn that on, it be beautiful, a beautiful sight."

He pulled away the last scraps of skin and held up a lump of white flesh the size and shape of a chicken breast.

"That be your dinner tonight," he said.

"Now I'll get one for you," I said, picking up my snorkeling gear.

"No," he said, "get another one for yourself. One with a pretty shell. Remember, you got to reach out for beauty."

In the fading afternoon light I swam back out through the turtle grass beds. I saw many more conch in the grass, and I was tempted to gather them up like Easter eggs, but I decided after all that one was plenty.

I was more interested in seeing if I could find Lobster Rock again. I couldn't. Though I lined myself up with the shoreline landmarks I had noted earlier, the coral head had disappeared. Now I noticed others like it, many others, all with the same general shape and size, but none with lobsters. I began to wonder if I had made it up, if it had been some kind of peculiar vision brought on by hyperventilating through my snorkel or by lying awake at night in my twisted sheets, my mind overstimulated with thoughts of swirling fish and ambling crustaceans. The more I thought about it, the more Lobster Rock seemed like some tantalizing childhood memory — something I thought I had seen, but perhaps really had not.

"Did you know that Hannibal invented trousers?" my dinner companion asked that night. His name was Earl, and he worked for the Federal Aviation Authority. He had come to Grand Turk to inspect a VORTAG antenna. "It's true. His elephants got cold crossing the Alps, so he put trousers on them."

"I saw eight hundred lobsters today," I said.

"No kidding. Where?"

"Out there somewhere." I gestured toward the dark water. "I doubt that I'll ever find it again."

The waitress brought my conch. It had been pounded flat and breaded, so it resembled that cornerstone of my Texas heritage, the chicken-fried steak. I suppressed a glum little wave of homesickness.

Earl knew about the elephant pants because his wife had been in a movie about Hannibal, *Jupiter's Darling,* starring Howard Keel. She was a former Weeki Wachi mermaid who used to work as a stunt double for Esther Williams. She could hold her breath for three minutes. Earl said that she almost died during the filming of *Jupiter's Darling* when, playing a woman pursued by Roman soldiers, she had to ride a horse over a seaside cliff and then dive down to seventy-five feet, where an air hose was waiting. But someone had forgotten to turn the air on, and she had passed out on the way to the surface.

Listening to Earl talk about his wife, I sank into a moody state of mind that was part lethargy, part longing. I wished I could hold my own breath for three minutes and dive without a scuba tank to seventy-five feet. I wished I had come to Grand Turk earlier in the year, in time for the manta rays. And I was irritated with myself for having misplaced Lobster Rock. If only I could find it again, I would go back out there now, in time to watch the lobsters creeping outward, one by one, on their nightly rounds on the moonlit floor of the lagoon.

12

Underwater Nights

Carrying my gear bag over my shoulder, I walked through the dark streets of Cockburn Town. A tinny instrumental version of "Lemon Tree" drifted through someone's open window, and up ahead, standing all alone in the street, a wild donkey opened its mouth and honked at me as if in anger.

In the dim moonlight I could make out an ancient, patched-together sloop moored just offshore, a merchant vessel carrying produce and charcoal from Haiti. Its mainmast was a tree trunk, and though it ran straight for most of its length it veered off near the top at a pronounced angle. Beyond the sloop, fishing boats patrolled the edge of the wall, their running lights coasting above the black water and their diesel engines throbbing with the seductive, fading-away moans of a locomotive.

I passed the Kittina Hotel and went down to the beach, where the crew of the *Fiesta* was loading tanks onto the boat for a night dive. *Fiesta* was a broad-beamed Island Hopper with a Bimini top and a dive platform at the stern. It belonged to Omega Divers, Grand Turk's only other dive operation. Omega was less of a shoestring operation than Mitch's Blue Water Divers. It sported two boats, a dive shop, and a steady clientele of more or less well-heeled guests from the Kittina. I had defected to Omega tonight because Mitch's compressor was broken, and in any case he preferred not to take his boat out on night dives; it carried so few passengers he could not make enough money to justify the inconvenience of getting cold and wet after a hard day's work.

I grabbed two tanks off the beach, waded into the waist-deep water, and hoisted them onto the boat. Eight or nine divers were already on board, wearing sweat shirts and warm-ups. They all knew each other and were in high spirits. One of the men sat in a captain's chair with a big Thermos in his lap.

"I am the custodian of the rum punch," he told me. "It will not be decanted until after the dive."

"Come on," a woman said, in a mock whine, "let us have some now."

"Elaine," he answered, "you are beautiful. But you may not have a rum punch at present. It is for *après*, goddammit."

Cecil Ingham, the Omega divemaster, came up over the stern and walked toward the wheel, his wet feet slapping the deck. His wetsuit, frayed and faded, was proudly festooned with a half-dozen patches documenting his various areas of diving expertise.

"Cecil!" one of the divers inquired heartily as the boat pulled away from the beach. "Any last-minute instructions?"

"Sure," he said, "don't go deeper than the bottom."

Cecil was a native Turks Islander who had gone off to college in Jamaica, majoring in mechanical technology. His calm, articulate air commanded the attention of this rowdy group. When he spoke, they grew suddenly serious and leaned forward to hear his quiet West Indian tones.

The *Fiesta* traveled along the margin of the wall beneath a sky full of stars in which the Milky Way shone and flickered like ocean phosphorescence. Cecil stood at the bow and shone a light on the surface, looking for the buoy that marked the mooring. When he found it, the crew tied the boat up, then lowered a down line off the stern, with a powerful blinking light attached that would help us find our way back to the boat. Silently, a little nervously, the divers began slipping on their tanks and weight belts and checking the beams of their underwater lights.

When I had suited up I looked over the side, watching the silent pulses of light — like heat lightning — that traveled outward from the light on the end of the down line ten feet below. Except for those periodic white bursts, the sea was black and threatening. Looking down at it, I remembered William Beebe's description of the ocean darkness he encountered when his bathysphere had descended to 1,000 feet: "The sun is defeated and color has gone forever, until a human at last penetrates and flashes a yellow electric ray into what has been jet black for two billion years."

For a long time I could not bring myself to dive at night. It seemed like such a somber, reckless activity, a journey not just into another element but into another dimension. Indeed, the divers on board tonight, as they shuffled in their fins toward the platform at the back of the boat, had taken on such a sudden air of gravity that they might have been preparing for a séance. I picked up my own fins and joined the procession. I had been on too many night dives by now to feel apprehensive, but my instincts still weighed against the idea of dropping into the ocean in the middle of the night. The first scuba divers who ventured into the water after dark were not even sure they would survive. Robert E. Schroeder, a night-diving pioneer, never left the boat without carrying a homemade "bangstick," a stainless steel barrel armed with .357 magnum cartridges that would discharge when rammed against the head of an attacking shark. Gradually, as they found that few sharks attacked, divers began to leave their bangsticks and their fears on the surface, until a night dive like this one could be considered no more than a mildly adventurous outing topped off with a rum punch.

I switched on the beam of my light, wrapped its loop around my wrist, and strode into the water, descending to the reef in a little cone of illumination. The ocean was alive with millions — billions — of orange and pink wriggling things a quarter of an inch long. I had no idea what they were — some sort of free-swimming larvae writhing about in the water with a blind, frantic energy. Mixed in with them were minute translucent shrimp, their backs arched and their many limbs beating furiously, propelling them through the vast ocean. When I shone the light on my slate to make a note, the pink wrigglies swarmed to it, covering the surface with their hopping, twisting bodies so thoroughly that I could not even see the white plastic beneath them.

Through this strange pink blizzard I made my way west toward the wall, swimming through the mounds and waving fronds of shallow-water coral. All around me the flashlight beams of the other divers swept back and forth, spotlighting or silhouetting features of the terrain. I had a vague recollection of some scene in an Italian movie I'd seen years ago, a scene in which a group of drunken house guests stagger through an ornamental garden in the dark with their flashlights, searching for a missing child.

I turned off my own light, wanting to savor the darkness. It fell like a cloak, but in only a moment my eyes had adjusted enough to be able to see shapes and shadows. Looking up through thirty feet of water, I could still see the stars, and all around me, as if I were surrounded by fireflies, bioluminescent fish blinked on and off with a bluish light.

I turned my light back on and swam over to a boulder of star coral to inspect the polyps that were lunging out of their cups, their tentacles waving in the current. At night, when the polyps are out, a growth of hard coral looks completely different — the hard, pleasing geometry of its surface obscured by innumerable buds of living flesh. When I looked close at these star polyps, I saw them feeding on the pink wrigglies, snatching them out of the water with invisible filaments and then drawing them down with their tentacles into their mouth slits. Caught in the polyps' mouths, the wrigglies struggled violently. Blind and utterly thoughtless, they were nevertheless impelled to fight for their lives. But it was a hopeless struggle. As I watched, the wrigglies slowly disappeared into the body cavities of the polyps, sinking as if into quicksand.

The polyps were not the only night hunters about. Cardinalfish and squirrelfish, their eyes as dark and fluid as compass bezels, swam through my light beam, and underneath every crevice I saw a skittering crustacean of some sort. Off in the distance I could make out Cecil in his worn blue wetsuit and specialty patches. With a group of divers crowded around him, he was snapping his fingers in front of a coral hole, trying like a snake charmer to lure a moray eel out into the open.

For all of the creatures that were on the prowl tonight, just as many were off duty, hidden away in their sleeping holes or wandering like sleepwalkers through the reef. Under a shallow ledge I came across a sleeping blue parrotfish. The fish had settled on the bottom as if it were weighted, but its fins did not move, and its eyes stared outward without taking anything in. With a finger I poked the parrotfish and rocked it back and forth, but it didn't wake. Gently I picked it up and held it, something I should not have done — both because it was rude and because my hands could disturb the protective sheen of mucus that covered its skin. I held the parrotfish at arm's length, turning it this way and that as if it were a plastic model instead of the real thing, then carefully put it back into its sleeping place.

One species of parrotfish, which I looked for but never found, encloses itself at night in a bubble of mucus spun from its mouth. Other fish sleep buried in the sand. Some lie on their sides, some hang listlessly in the open water. I encountered a small filefish that hovered blankly above the seabed with its nose pointed downward, as if it were contemplating the grains of sand at the bottom of the reef. When I touched its tail, it did not move away in response but literally fell, bouncing on its nose in the sand. Nearby, another filefish was stuck as if by adhesive to the latticed blade of a sea fan, and showed no concern at all when I touched its rough, warty skin. Next a somnambular procession of doctorfish swam between my legs. They moved in single file, and though they managed to hold their ranks they appeared dazed and directionless. Ahead of them a queen triggerfish snoozed in the open water and did not even notice when the lead doctor woozily plowed into it headfirst.

I smiled behind my regulator at the idea of all these zoned-out creatures bumping into each other in the darkness. This comic reverie took the edge off my lingering apprehensions about night diving. But when I saw the other divers begin to ascend, their light beams drifting up to the flashing star suspended from the down line, I began to notice again the black immensity of the world surrounding me, a world in which I was alone and unwelcomed, and which neither my knowledge nor my imagination was powerful enough to truly comprehend. Rising to the surface to join those congregating beams of light, I turned in slow spirals, not wanting to turn my back on whatever might be out there.

But I knew that the true dangers of diving, whether at night or in the daytime, are far more pedestrian. The great hazards are inattention and poor planning: a tank inadvertently drained of air, the extra thirty feet of depth that clouds your judgment, the aimless underwater itinerary that leaves you lost and far from the boat, the petty annoyances that can lead to panic.

Just such an annoyance waited for me at twenty feet — a layer of sea wasps, small stinging jellyfish that swim near the surface after dark and are attracted by beams of light. They attacked me on my bare shoulders and arms and then on my chest, slipping beneath the fabric of my farmer-John wetsuit. The stings were deep and penetrating. Like coral polyps, sea wasps subdue their prey with the venomous trip wires known as nematocysts, and I could feel these invisible threads tattooing my skin with their poison. Sea wasps are painful but relatively harmless, unlike the box jellyfishes of the South Pacific, which discharge a toxin powerful enough to kill an adult. Nevertheless, I felt as if I had been attacked by a swarm of bees, and I quickly descended to get below them. When I headed up again, I was more careful, checking to make sure that none of their milky, translucent shapes were visible in the beam of my light.

When I slipped through the surface I found myself facing

the lights of Cockburn Town a hundred yards away. The other divers were climbing onto the boat, and I waited my turn, resting on the surface with my buoyancy compensator inflated and my skin still tingling with pain. I closed my eyes and leaned my head back, resting my hands on my chest like an otter. The water was still and cool, and the sea wasp stings slowly faded in intensity until their effect was almost pleasant. When it was my turn I reluctantly kicked over to the diving platform, took off my fins, and climbed the ladder, my tank and weight belt suddenly reminding me of the curse of gravity.

The guardian of the Thermos was already on board, wearing his Scubapro warm-up jacket, a towel wrapped around his head.

"See anything good?" he called to the diver ahead of me.

"A little nurse shark," he said, dropping his weight belt and wiping his runny nose on his neoprene sleeve. "A moray eel. Some other shit I can't remember. *Now* may I have a rum punch?"

During the week that Mitch's compressor was down, I did my diving from the *Fiesta* or roamed around the island on my scooter looking for new places to snorkel. One evening I went with Matt, the Smithsonian botanist, to a place called White Sands, near the southern tip of the island. It was an abbreviated patch of beach and coral rock, with a single picnic table beneath a shedding canopy of casuarina trees. It was dusk when we got there, and the air was still. We had not been out of the cab of the Smithsonian truck for thirty seconds before we found ourselves in a state of unbearable agitation, so covered with mosquitoes that when we slapped our skin to kill one we killed two dozen and coated our palms with a black smudge of mosquito bodies.

"Let's get out of —" I started to say, but then closed my mouth as the insects swarmed into it. I spit out as many as I could and joined Matt in running back to the truck, where we rolled up the windows and slapped the mosquitoes off our skin until we finally settled down.

Matt was laughing an exhilarated laugh, as if we had just gone for a bracing swim in polar seas.

"They can't get us underwater," he said. "Ready? One . . . two . . ."

We ran out of the truck with our masks in place on our faces, holding our fins in our hands. By the time we hit the water, the intolerable sensation of the mosquitoes swarming over my skin had almost driven me mad — it felt like an animate hair shirt. I dove down to the bottom and coasted along inches above the sand, shuddering like a fly-covered horse. When I came up for air and took a deep breath through my snorkel, I inhaled a mouthful of mosquitoes.

The situation grew more tolerable as we swam farther from shore, and finally I was able to forget about the plague on the surface and concentrate on what was happening below. It was near dark. The coral polyps had not yet come out, but I could detect a sense of growing urgency and excitement on the reef as various kinds of fish began their search for prey or nighttime shelter. All around me fish were swarming, schooling, nipping furiously at each other — as if the darkness brought with it a surfeit of nervous energy. A big stoplight parrotfish — a terminal male — streaked across my field of vision with astonishing velocity, pursued by another, even larger fish of the same species and station. And in the distance a bar jack, powered by lashing strokes of its narrow body, dove into a cloud of silversides, missed its target, and almost slammed into the sand. When it grew darker I watched another stoplight parrotfish scoop out a kind of nest in the sand and then repose there like an odalisque, a faraway look in its already faraway-looking eyes.

Matt snatched a lobster that was just venturing out from its crevice, but when he saw that it was undersized he let it go. In its panic to get away, the lobster flew through the water, bending in on itself and moving backward with jerky flaps of its tail, its antennae trailing. After that, Matt scared up an eagle ray and swam behind it, flapping his arms in mimicry as it soared over the reef.

Matt was the ideal diving buddy: skillful, high-spirited, game, uncomplaining. In his off-hours from the Smithsonian, he was pursuing his divemaster certification under Cecil's tutelage, so he was often on board the *Fiesta*.

The days were clear and windy now, an "outwind" — as Turks Islanders call a west wind — driving the surf against the beaches and sea walls. Loading the *Fiesta* from the beach became a test of endurance, as we carried our tanks into the surf, bracing ourselves against the undertow, which threatened to throw us off balance and slam our ankles against submerged rocks, while the idling motors of the boat spewed diesel fumes into our faces. In the high seas we had to be careful at the end of a dive, when we approached the bouncing platform at the stern of the boat. Twenty years earlier, as a college student, I had visited a fortune-teller who had warned me about this specific hazard.

"You are a gifted pianist," she had told me as she held my hand and gazed at the lines in my palm.

"Well, no," I had answered, remembering a nun with hairy knuckles who had sighed with disgust at my inability to master even a measure of the "Volga Boat Song."

Undeterred, Madame Hipple peered deeper into my palm. "Watch out!" she said. "I see you swimming under the water. You are coming up for air, and there is a raft or something made out of wood that you do not see. Unless you are careful, you'll hit your head on it and drown."

Two years later I went on my first open-water dive in the Gulf, and when I looked up on my return to the surface I saw the wooden diving platform slamming up and down above my head. The accuracy and aptness of Madame Hipple's warning startled me, and over the decades, diving in rough seas, I've carried that prediction like a talisman.

At night the wind was usually low, and the seas calm, but I never returned to the *Fiesta* without taking a hard look at the underside of the dive platform. The rum punch crew were long gone, back to their jobs as hospital administrators and utility company employees in the Midwest. Now there was Jerry, the marina owner from Toronto, who used his diving skills back home to retrieve sunken snowmobiles from under the ice; Cliff, the contractor from Newport Beach, California, who had flown fighters with John Glenn in Korea; Eric, a banker from England; Frank, a Dutch water engineer who had come to Grand Turk to recommend ways to improve the water supply ("I'm for more roof catchment. A cistern for every house!"); and Carl, a doctor from Indianapolis who had visited, it appeared, every other diving site in the world.

"Look what happened to me in the Caymans," Carl said to the other divers by way of introduction. He held out his right hand, which had a half-moon scar in the space between the thumb and forefinger. "Grouper bite. It was supposed to be a tame fish. We were feeding it pieces of bread, but it decided to chomp down on my hand instead. You know, they say if a grouper bites you, you're supposed to keep your hand still and let him take his teeth out and back off." Carl looked down sadly at his hand. "That's not easy to do."

One night we dove beneath a bright moon. The pink wrigglies were gone, but the coral polyps fanned the water with the same hunger as before, drawing in the nearly invisible feast of plankton that drifted over the reef. When Matt and I turned off our lights and cruised along the edge of the wall at forty feet, the moonlight was strong enough for us to see each other from thirty feet away. Swimming along in the soft darkness, I saw the light from bioluminescent plankton sparking off my hands. It gleamed and twinkled like pixie dust, following my every move. I thrust out my hand, and the glowing plankton flew from my fingertips as if I were a farmer scattering magic seed. When I looked back at Matt, I saw the contours of his body outlined in light. And if I narrowed my eyes into a squint, I could not see his body at all, only the bright aura that enclosed him.

Exhilarated, I dove deeper, descending along the face of the wall at a sharp angle, my outstretched hands throwing up a spray of light. It seemed like a miraculous phenomenon, but in fact bioluminescence is so common among marine creatures as to be ordinary. The bodies of squids and jellyfish are often outlined in pearly beads of light, and certain fish — particularly those found in the black depths of the ocean — have light-producing organs below their eyes that act as headlamps, illuminating the dark crevices where they search for food. In most luminescent creatures the light comes from specialized cells called photophores or from glowing bacteria harbored within the body tissues. In some cases the light blinks on and off as if produced by an electric circuit — the result of a compound called luciferin coming into contact with an enzyme called luciferase. But bacteria tend to emit light in a steady gleam, so to turn it on and off some creatures have developed retractable flaps of skin.

These millions of glowing plankton animals did not form

an illuminating whole. The light was scattered and brief, a tinsel light that melted back into the darkness after its instantaneous existence. But even at sixty feet the moonlight was still bright enough for me to see the contours of the reef and the moving shadow shapes of fish. Cruising along a coral shelf, I noticed a thick, gray, weaving shape, and when I turned my light back on I saw what looked like a sea dragon: a green moray eel five feet long, lounging out in the open with its body draped over the coral and its fierce eyes shining in the beam of my light.

It was the biggest eel I had ever seen, almost as thick around as my leg. When I approached, the moray did not even bother to retreat but just looked at me with its beady eyes, set so far forward on its narrow head that they almost touched in front of the siphonlike nostrils. It opened and shut its jaws as if in threat, but I knew that this was just the way it breathed, swallowing water in order to force it back toward its diminutive gills. When the eel's mouth was open, I could see sharp, pale teeth lining its jaws. A moray's teeth slant backward, allowing it to hold its prey like a Chinese finger trap as the victim tries to wriggle free. Those teeth are the main reason divers are advised not to stick their hands into coral holes. The common assumption about morays is that they lie in wait in these holes, waiting for a fish to swim by, and then spring out like a striking rattlesnake to grab it. In fact, they often hunt on the move, wending through the coral at night to seize fish or mollusks out in the open or slinking into narrow holes to pluck them out of their hiding places. Morays have a blundering, bullying appearance, but they are efficient and rather elegant predators. They have been observed to eat octopuses an arm at a time, twisting the creatures around in their mouths and wrenching off each limb in turn.

I couldn't take my eyes off this moray — its worn green skin, its wattled old man's neck. Humans have always had a charged fascination with moray eels. The Romans apparently revered them and sometimes kept them by the thousands in household salt water ponds renewed by channels cut through to the sea. There are far-fetched accounts of slaves being thrown into the ponds and devoured by the eels for the entertainment of dinner guests, but more intriguing are the stories of almost surreal infatuation — the rich young woman who lovingly outfitted her moray with earrings, the noble who could not bear to sell his private stock of morays to Julius Caesar, or the powerful politician who was publicly rebuked in the Senate for moping about the death of his beloved eel.

I backpedaled against the current so I could hold my position in front of this immense moray. I kept a discreet distance between us — not out of fear but out of courtesy. To a human eye, green morays are repellent, even demonic. They seem to have materialized out of some primal nightmare. But they are curious and, to a degree, even sociable fish. At high-traffic dive sites, where they have grown accustomed to receiving handouts from humans, they can often be lured out of their holes. Once on a night dive I had been instructed by the divemaster to stop in front of a crevice and hold out my closed fist. Slowly, a big moray waggled out and touched its blunt nose to my fingers. I held my breath, thinking of the damage those jaws could do to my hand if the eel's mood changed, but it just kept swaying, and after a moment I grew bold enough to stroke its slimy flanks.

I did not reach out to this eel tonight. I did not want to ruin it. I felt bad about thoughtlessly handling that sleeping parrotfish earlier and worried that I was in danger of making an underwater pest of myself. Part of me, it was true, wanted to insinuate myself into the lives of the creatures on the reef — to touch them, incite them, ride them, make them pay attention to me. But every time I gave in to this impulse I felt shut out from the true life of the reef and sealed once again into my outsider's bubble.

When I turned off the light, the moray began to move, a colorless form slinking through the coral, phosphorescent sparkles bouncing off its head. I took my slate out of my buoyancy compensator pocket and wrote: "When the moon hits your eye like a big pizza pie, that's a moray." Shining the light on the slate, I chortled so hard at my own humor that a burst of bubbles escaped my regulator with enough force to create yet another flurry of light.

13

Lord Face of Water

In the stillness of the morning the water was so calm and polished that driving a boat across it seemed an act of violence. Mitch's outboard cut an ice-blue wake behind us, and the deep V-shaped hull glided along the surface with the smooth velocity of a toboggan. When we stopped at the buoy above Harmonium Point the wake quickly dissolved, the noise and the exhaust drifted away, and the water lay so still and silent I imagined that I could bounce a penny off it.

Looking down through sixty feet of water, I saw moving shapes that were as tantalizingly indistinct as spirit forms glimpsed through a crystal ball. I wondered what it would be like if the world were tilted on its side, with the ocean rising up beside me like a curtain that I could walk through, rather than plunging down into it. "How much more curious about [the sea's] unfamiliar creatures many of us might be," wrote the zoologist Alister C. Hardy, "if the sea were in fact separated from us by a vertical screen — over the garden wall as it were — instead of lying beneath us under a watery floor. Who as a child has not envied the Israelites as they passed through the Red Sea as if marching into a continuous aquarium: 'and the waters were a wall unto them on their right hand, and on their left'? What might they not have seen?" I remember envying the Israelites myself, but to roam freely in this lovely water seemed more miraculous to me now than anything they might have seen from the bed of the Red Sea. In silence, as thoughtfully as if we were dressing for battle, the divers on board began to gear up. These were practiced motions for me, and I was fussy about their order. First I pulled on my farmer-john wetsuit, smoothing out the thick neoprene wrinkles until the tips of my fingers were sore. Then I strapped my knife to the inside of my calf, and put on the weight belt with its sixteen pounds of crude gray weights. I seated the first stage of my regulator on the tank valve, screwed it tight, then attached the inflator hose to the buoyancy compensator. I arranged the various components of the regulator the way I liked them, the primary mouthpiece close at hand, the octopus tucked out of the way and held in place by the Velcro tabs on the b.c.

When I twisted open the tank valve, I could hear the air flowing through the hoses, and I automatically put my ear up against the valve to listen for leaks, then took a test breath from the regulator mouthpiece. The compressed air was dry and harsh, an insistent rush that seemed to ricochet off the back of my throat like a solid object and then drop down to my lungs with the force of gravity. According to my submersible pressure gauge, the tank had a good fill — almost 3,000 pounds per square inch, enough air to stay down for an hour or so if I didn't dive too deep. I set the tank and the b.c. on the gunwale of the boat, squatted down to slip my arms through the straps, then fastened the b.c. with its various plastic buckles and its Velcro cummerbund.

All that remained was to check that no hoses were tangled in the b.c. straps, to slip on my fins and spit in my mask to keep it from fogging, and to make a few last-minute adjustments. Knowing that the pressure of deep water would compress my body, I tightened the weight belt as much as I could to ensure that it would not be loose underwater. With my dive knife I sharpened the point of my carpenter's pencil, then scrubbed the surface of my slate with a big eraser to clear it of the previous dive's notes. With the slate and the pencil stashed in one of the pockets of my b.c., I pulled the mask over my face with the flourish of a knight slamming down his visor. The prescription lens made the world sharp and alluring. For a moment longer I sat there, absorbing the warm sun on my forehead, enjoying the clarity of my vision. When I felt it was time, I put the regulator in my mouth, set the timer on my watch, and rolled over backward into the water, the heavy tank on my back bursting through the surface like a depth charge.

Through trial and error I knew just how much weight I should wear on my belt to drift downward without bobbing back to the surface or breaking the rhythm of my entry into the water. Rolling backward off the boat and sinking to the bottom became a single continuous motion, and it gave me a kind of Zen pleasure to perform it well, to disappear from sight in a silent veil of bubbles.

Most often I descended feet first, my ears clearing automatically as I dropped like a parachutist to the rubbly coral floor. Today I looked upward as I sank, watching my bubbles grow in volume and velocity as they rose to the surface. Except for the bubbles, there was not a blemish — not a wrinkle — in the transparent ceiling overhead. As I fell toward the bottom, the top of the ocean grew larger in my sight, until it seemed like the absolute summit of the universe, the outer margin of what could possibly be imagined or perceived. Gazing at that satiny blue frontier, I recalled the name of an Aztec god I had once come across in a book about the conquest of Mexico: Lord Face of Water. Just the name sent a thrill of reverence through me. A childhood of intense Catholic indoctrination had long ago ruined me for belief in God, but now for the first time in my adult life I had a notion that behind the implacable facade of nature there might reside a divine personage of some sort — not good, not bad, just brooding and eternal, keeping watch at the places where different worlds conjoin. And I thought if I were ever to worship anything again, it would be Lord Face of Water.

I drifted down to the top of the wall and hovered just above the growth of star coral at the edge of Harmonium Point. The open sea in front of me was featureless, without demarcations of any sort. No fish were present to give it scale or depth of field, so it seemed flat, an empty blue screen waiting for something to be projected upon it. I heard a muffled explosion and looked up to see the another diver hitting the water. Above him, beyond the glassy surface, other divers were preparing to go down. I watched them sitting on the edge of the boat, checking their masks, putting their regulators into their mouths, arranging their tendrils of hoses. One by one they slipped over, their motions awkward as the heavy yellow tanks pulled them down through the air and slammed them into the water. But from below their passage into the ocean looked as graceful and clean as if they were falling into a mirror.

The divers congregated on the bottom, and then Mitch came down to lead them through the shallow grooves and canyons to the face of the wall. When they swam past me, I joined them, bringing up the rear. We swooped down below the promontory, diving deep along a section of the wall covered with red algae, then below an overhang festooned with a few precious sprigs of black coral. Among divers, black coral has an overblown mystique. Though more closely related to the reef-building corals than to the soft gorgonians, black coral has a wavy, downy appearance. But its stems are tough and tensile, made up of protein secreted by the polyps that cover them. When dried and polished, black coral is as dark and glossy as ebony, and because there is such a demand for the jewelry made from it, the coral has been ravaged in the wild. It grows best at depths of sixty feet or more, sheltered beneath overhangs and out of the path of direct sunlight. It was once common at relatively shallow depths, but not anymore. The most easily accessible specimens have been stripped from the reefs, and it is now a rare sight. Divers have to go deeper and deeper to find it, and once it is harvested it takes ages to grow back. For some divers it has become a test of macho prowess to plummet down into the darkness and return with a branch of black coral as a souvenir of the abyss. The trinket shops of Cozumel and Grand Cayman are filled with necklaces and key chains and klutzy figurines carved from black coral that has been pried out of the reef at depths of 250 feet and hauled up, along with its dying polyps, into the light. From glancing at these objects I had always had the impression that black coral grew in immense clumps, dark as coal and hard as petrified wood. But these corals were small and had a willowish softness. They waved in the faint current, and except for their slim black stems they were white.

The familiar tour continued along the face of the wall. Below us the reef shelved off into blackness. I dropped down ten feet below the other divers and swam on my back, helping Mitch keep an eye on them and at the same time studying the ridge line of the wall, the soft waving corals backlit by the blue sky far above. A line of doctorfish swam by overhead, their deep gill slits giving them the look of rodents with overstuffed cheek pouches. All along the wall, the fanlike bristles of segmented worms popped back into their burrows as we passed. There were many species — fanworms, feather duster worms, Christmas tree worms — feathery spirals and tiers radiating outward from the coral rock, as fragile-looking as dandelions. Some were slower to retract than others, but most were gone in an instant as our shadows fell across them. These bouquets are feeding tentacles whose job is to filter nutrients from the water and pass it down to the unseen creature residing deep in its hardened tube in the coral. When the tentacles blinked out of sight, they did so with the swiftness of a magic trick, but they reemerged warily, in slow motion, the way a flower unfolds in a time-lapse film.

Beneath an overhang we came across a swarm of the small blue-and-yellow fish known as fairy basslets. They were skittish fish, and when I got too close they would dive into their holes with confounding speed. But they did not seem to mind being watched from a few feet away, and I noticed that at any given moment a number of them would be swimming upside down with perfect ease. It struck me as marvelous that their orientation in the sea was so casual, that their world appeared to have no top or bottom. Inspired, I hung upside down myself, my fins pointing toward the surface, my eyes looking down toward the gloomy depths.

At rare moments while diving over the years, I have lost track of up and down. It happened once on a deep dive in a murky lake, when I descended into a layer of swirling black mud, and once in the open ocean amid the girders of a sunken drilling platform that stretched off in either direction into an undifferentiated haze. In both instances I felt a smothering sense of being lost and had to remind myself to stop and study the upward drift of my air bubbles in order to find the way home.

In this clear water there was no question about up and down. In addition to the obvious visual cues, my regulator was harder to breathe through upside-down, and I could feel the blood beginning to pool in my head. Nevertheless, I closed my eyes and sailed along, inverted, in the current. One of the great joys of diving is discovering how much more *spacious* the underwater world is than the terrestrial one. Walking on land, you always have that enormous unusable sky above you. You are confined, hounded by gravity. You walk into your house through a ground-level door, you stand on a chair to change a light bulb, you climb a ladder to brush away a spider web in the corner of the ceiling. But flood the house with water and your world expands. All that fallow space above your head is suddenly yours — you are able not only to reach it but to *inhabit* it. There is more room in the world.

When I grew tired of hanging upside down I wheeled back into the horizontal plane. I ducked into a canyon and floated up a coral slope, the sun growing stronger and my body more buoyant as I emerged from the deep water. Rising over the reef summit, I felt delirious with the range of my movement. Within the limits of physiology, I could go anywhere — sink into a crevice, soar along a coral pinnacle, wend my way through a tunnel with ghostly precision, not having to worry just yet about returning to the surface for air.

I was lucky to know this freedom. Though I resented a bit the ungainly air tank on my back and the various cumbersome accessories that reminded me how imperfectly adapted I was for this environment, I still felt amazingly fortunate to have been born at a time in history when I could take advantage of these things. For ages, inventors had dreamed of creating a machine that would allow people to swim underwater for sustained periods of time, but it was not until my own generation that this dream became a commonplace reality.

The idea of a self-contained underwater breathing apparatus has been brewing in the human imagination for thousands of years. An Assyrian relief carving dating from 885 B.C. provides the first known depiction of scuba divers. Wearing headbands to keep their flowing hair out of their faces, they coast along through stylized curlicues representing the sea. Their diving suit is a kind of choir robe, and their equipment consists of an inflated leather pouch that they hug to their chests. The divers suck air from the pouch by means of a short tube. They wear belts and backpacks of some sort — possibly filled with weights to counteract the formidable buoyancy of the air bags.

It might seem, at first glance, that the Assyrians had solved the problem of breathing underwater. They simply took an air supply down with them — perhaps after squeezing air into the sack with a bellows. But if you look closely at the carving you see that they are swimming only a few feet below the surface. They must have been disappointed when they tried to go deeper and could draw no air from their leather pouches. They would have learned from trial and error what seventeenth-century science confirmed: that there is a consistent spoiler to underwater exploration known as pressure physics. It works like this: if you are standing on land, the force of the atmosphere exerts pressure upon your body of 14.7 pounds per square inch. But water weighs more than air. When you jump into the ocean and swim downward, the density of the water causes the atmospheric pressure to multiply. At just thirty-three feet, you are under the equivalent of two atmospheres; at sixty-six feet, three; at ninety-nine feet, four. This has many effects, the most immediate being that an increase in atmospheric pressure causes gas — such as the air in human lungs — to compress, to assume a smaller volume. Drag one of those Assyrian air pouches down to thirty-three feet or so, and there is simply less air in it to breathe. Not that it matters so much, since the pressure of the water has compressed the diver's lungs as well. At that depth they are too weak to draw the remaining air in the pouches through the breathing tube.

It remained more practical for the professional divers of antiquity to abandon such hopeful devices and simply hold their breath underwater. The idea of diving as a means of gaining military advantage or economic gain took root early in the clear waters of the Mediterranean Sea. Professional divers in ancient times ventured beneath the surface to gather ornamental coral and mother-of-pearl. They dove for real pearls as well, and for the murex snails from which purple dye was manufactured. Sponge diving, we know from Oppian and Aristotle, was a thriving but miserable occupation. It appears that sponge divers slit their nostrils — or had them slit by their employers — perhaps in hopes that they could profitably mimic the gill respiration of fish. Holding their breath, they plunged toward the bottom — eighty or ninety feet down — with the aid of a heavy stone, and without the benefit of masks or goggles to keep their vision clear. In order to see better, they filled their mouths with oil and spit it out on the sea floor a little at a time. The oil floated upward, where it made a smooth patch on the surface that allowed more light to penetrate the depths. Arab divers used this technique, and according to al-Mas'ūdi, a tenth-century naturalist, they also pierced their eardrums (presumably in an attempt to bypass the problem of ear squeeze) and communicated with each other underwater by making sounds similar to the barking of dogs. Pearl divers in the Persian Gulf later developed goggles with lenses of transparent, wafer-thin tortoiseshell.

Herodotus skeptically passes on the story of a Chalcidian diver named Scyllias who was employed by the Persian army during the Median wars but defected to their Greek enemies by swimming nine miles underwater. Scyllias and his daughter, Cyana, later made a commando raid on the Persian fleet, diving below the waves to cut the ships' anchor lines and cast them adrift in a fierce storm. In gratitude the Greeks erected statues of the two divers in the Temple of Diana.

A durable legend holds that Alexander the Great made an underwater pilgrimage in a cage made of glass, through which he observed the passing sea life, including a creature so gigantic that it took four days and nights to swim past. Fable aside, Alexander's cage sounds like a diving bell, an instrument used on occasion by Aristotle's sponge fishermen. The theory of a diving bell is simple — a large vase, jar, or tub is turned upside down and forced downward through the water. The air trapped inside is compressed by the increased pressure of the water and driven to the top of the container. Within this shelter a person can breathe — though the greater the depth, the more compressed the air becomes and the less is available to breathe. After a while, of course, the air turns bad, and if it cannot be replenished somehow, the diver must surface. By the seventeenth-century the diving bell was more or less perfected. Using natural light or candles for illumination, the divers sat fully clothed and dry on benches, opening taps to let in reserves of fresh air when the atmosphere grew close.

Leonardo da Vinci devised several impracticable diving apparatuses, including a full-faced leather helmet with glass eye holes and a series of spikes radiating out from the head. The spikes, Leonardo explained, were to protect the diver from fish. The inventor imagined that the occupant of his helmet would be able to breathe by means of a long snorkel extending all the way to the surface. This certainly would not work, because the water pressure at more than a few feet would not allow the diver's lungs to draw breath from the surface. Leonardo had better luck with footwear. He designed fins and also a prototype of the webbed gloves that have cropped up now and again, without much staying power, in the underwater fashion parade.

All sorts of weird devices came and went: large-bore snorkels the size of portable chimneys; watertight suits crowned with long, flowing topknots whose open ends rose above the surface to collect air; robot machines in which the diver wore a pair of special leather shorts and encased his torso and head in a metal cask with eerie eyeholes. These outfits look ridiculous to us today, but at the time many of them were taken seriously and, after a fashion, they worked quite well. "These machines," wrote a French journalist of one design in 1797, "incorporate so many advantages that it would appear impossible to improve on them."

In 1819 a man named August Siebe had great success in applying the principle of the diving bell to individual use. He invented the forerunner of the familiar deep-sea diving suit, with its weighted shoes, metal helmet, canvas coveralls, and pressurized air hose tethering it to the surface. A sophisticated diving suit of this sort enabled a diver to go much deeper than was possible before and for much longer periods, but he dropped through the water like a plumb bob. He did not swim, he sank. And he was as much a hostage to surface air as if he were breathing through a hollow reed.

The trick was to design a practical machine that would allow a diver to carry his air with him. Attempts to do so involved the use of underwater bellows, tanks of compressed air worn around the waist like inner tubes, or billowy air bags attached to a hand-cranked portable compressor. For the most part these designs were dead ends, but in the 1860s two French inventors, Benoit Rouquayrol and Auguste Denayrouze, created a device they called the *aerophore*. The *aerophore* consisted of two main components — an air-filled canister about the size and shape of a lawn mower engine, which the diver wore on his back, and a new creation, a "regulator," which automatically adjusted the pressure of the air in the tank to that of the water surrounding it. A hose ran from the tank to a surface air supply, but the diver had the option of disconnecting the hose and plodding about for a few moments using only the air in the tank. Later versions of the Rouquayrol-Denayrouze apparatus came with a helmet, but in the earliest model the diver breathed through a flexible tube with a mouthpiece. An illustration from the period shows an *aerophore* diver wearing a kind of wetsuit and what look like sandals. Except for a noseclip, he is bare-faced. The inventors recommended against goggles. Though the diver would be virtually blind without them, Denayrouze was of the opinion that "the action of sea water on the eyes is more of a tonic than otherwise."

Jules Verne used the Rouquayrol-Denayrouze invention as the basis for the diving outfits worn by Captain Nemo's men in *20,000 Leagues under the Sea*. The *aerophore* was a tantalizing invention, but its hour had not yet come. The primitive regulator worked well enough, but in order to provide more than a few minutes of freedom underwater, the air in the tank would have to be compressed to a pressure that the metal of the time could not withstand. And the whole ensemble, with its helmet and weighted shoes, showed a limited grasp of the possibilities of the underwater environment. Rouquayrol and Denayrouze, like Siebe and many others before them, could not seem to get away from the idea that divers should *walk* underwater, that they should trudge along with heavy footsteps as if they were making their way through a storm on the earth. All that water between the surface and the sea bottom — the water in which modern divers fly like birds — was for these pioneers merely a deadly and lugubrious atmosphere. When a diver plunged into the sea, he did not think about flying; he thought about where he would set his feet.

Other inventors were driven by the longing to soar underwater. "To swim fishlike, horizontally," Jacques Cousteau writes in *The Silent World* of his first test dive of the Aqua-Lung, "was the logical method in a medium eight hundred times denser than air. To halt and hang attached to nothing, no lines or air pipe to the surface, was a dream. At night I had often had visions of flying by extending my arms as wings. Now I flew without wings. (Since that first aqualung flight, I have never had a dream of flying.)"

On the eve of World War II, Jacques Yves Cousteau was a young artillery officer in the French navy stationed in the Mediterranean port of Toulon. He had a restless, tinkering intelligence and had been an impassioned filmmaker since he was a teenager, directing a series of melodramatic shorts in which he usually cast himself as the villain. By the time he was assigned to Toulon, he had been around the world on a training ship and had noted with interest South Sea pearl divers and Indochinese fishermen who dove beneath the surface and captured sleeping fish in their bare hands. A car accident while he was a student at the naval aviation school had left him with a slightly crippled right arm that ruined his plans to be a pilot.

But his lame arm did not prevent him from taking up the new sport that was being born in those years along the coast of France between Nice and Marseilles. Before skin diving had a name, it was known variously as underwater swimming, underwater hunting, goggling, or — for its assaults on submerged landforms — "inverted alpinism." Its adherents had grown impatient waiting for technology to come up with the proper underwater breathing machine and had decided to turn themselves into "menfish" by lung power alone. The boom in skin diving was sparked, however, by other inventions, most notably the face mask marketed by a Russian emigré named Alec Kramarenko in 1937. Before that time the basic underwater vision aid had been goggles like those worn by Japanese pearl divers. Until 1936, however, when diving goggles were first manufactured, aspiring divers had to make their own. Guy Gilpatric, the American author who is generally regarded as the founder of skin diving, used aviation goggles, which he waterproofed with putty. Others built their own from scratch out of window glass, inner tubes, cow horn.

Kramarenko spent years refining the face mask that would help make goggles obsolete for diving. Pierre de Latil, the author of the authoritative *Man and the Underwater World*, recalled visiting the inventor in his workshop. "One day he showed me his 'museum,' an old wooden case full of his various models, and I was astounded to see the amount of painstaking labour and almost superhuman patience involved in the work of producing something so simple — or should I say something that now seems so simple — as a satisfactory underwater mask for divers."

The separate lenses of goggles tended to create a double image, a problem Kramarenko solved by using a single pane of glass. Unfortunately, this new design trapped more air against the diver's face, and as he descended, the increasing water pressure created a painful squeeze. It took another inventor, Maxime Forjot, to solve this problem. Forjot created a mask that covered not only the eyes but the nose as well. This allowed the diver to exhale through his nose and break the vacuum lock of the trapped air.

Cousteau had had a moment of epiphany the first time he ventured underwater in a Mediterranean bay in 1936 with a pair of goggles. "Sometimes we are lucky enough to know that our lives have changed," he wrote, "to discard the old, embrace the new, and run headlong down an immutable course. It happened to me at Le Mourillon on that summer's day, when my eyes were opened on the sea."

The invention of the diving mask brought the underwater world into ever clearer focus. At the same time Forjot patented the snorkel, a short breathing tube that allowed a diver to swim along on the surface with his head submerged so that he could scan the seascape without lifting his head to take in air. And a few years earlier Commander de Corlieu had patented a workable version of the foot-fin that Leonardo da Vinci had foreseen in the fifteenth-century. These flippers added considerably to the power of a diver's leg strokes, extending his underwater range at the same time the mask and snorkel enhanced his vision and his staying power. With these tools, and with homemade harpoon guns and spears, the manfish became an astonishingly successful underwater predator. The fish were so innocent about this new threat that divers could swim up to them and, with the flourish of a fencing master, simply stab them.

By the time masks, fins, and snorkels became commercially available, Cousteau had been a manfish for several years. He dove with a physics professor's son named Frédéric Dumas, who impaled fish with sharpened curtain rods. Another of his companions was a fellow naval officer, Philippe Tailliez. The three of them ranged through the coastal waters, spearing fish, snorkeling with breathing tubes fashioned from garden hoses, and making primitive underwater movies. Cousteau's curiosity reached farther than that of most of his fellow Mediterranean fish killers. By temperament he was more an explorer than a hunter, and the limits of breath holding were a great frustration. "When we had attained the zone of sponge divers," he writes, "we had no particular sense of satisfaction, because the sea concealed enigmas that we could only glimpse in lightning dives. We wanted breathing equipment, not so much to go deeper, but to stay longer, simply to live a while in the new world."

Twice Cousteau almost killed himself experimenting with an oxygen rebreathing system that the gunsmith of his ship built to his specifications. The oxygen rebreather had been around since 1876, when its inventor, Henry Fleuss, was hauled up out of the water vomiting blood after testing the apparatus. The idea behind the invention had a beguiling simplicity. Instead of swimming around with a limited and cumbersome supply of compressed air containing the same mixture of nitrogen and oxygen he breathed at the surface, a diver would breathe a small amount of pure oxygen, which could be recirculated through a scrubber. Since the oxygen was continuously recycled and purified, the diver did not run out of air and could stay down as long as he pleased. And it was silent and self-contained, with no noisy bursts of exhaust bubbles. The only problem is that beyond about one atmosphere of depth, or thirty-three feet, pure oxygen turns toxic. At forty-five feet Cousteau began to fade out, but he managed to jettison his weight belt before he rose unconscious to the surface. When he tried again the next summer, the same thing happened.

He also tried the invention that Commandant Yves Le Prieur had showcased at the 1937 Paris fair in an underwater production called *L'Aquarium Humain*. The Le Prieur diving lung was relatively efficient but unwieldy. The diver walked along the bottom with a compressed air cylinder attached to his chest and from time to time opened a valve to deliver air into a mask that covered not only his eyes and nose but his mouth as well.

Cousteau found the Le Prieur apparatus a nuisance to operate. "Instead of Le Prieur's hand valve, I wanted an automatic device that would release air to the diver without his thinking about it, something like the demand system used in the oxygen masks of high-altitude flyers."

War had come by this time. Cousteau, whose brother Pierre was a leading propagandist of the Vichy government, joined the Resistance, using his diving skills to scout out Italian shore installations. Food was scarce, but Cousteau and the other underwater hunters managed to supply their families with a bounty of seafood.

Amid all this wartime intrigue and peril, Cousteau traveled to Paris in 1942 to meet Emile Gagnan, an industrial gas expert who worked for a company called Air Liquide, one of whose executives was Cousteau's father-in-law. When Cousteau described for Gagnan the self-adjusting regulator he had in mind, Gagnan retrieved a gadget from his cluttered office and said, "Something like this?" In occupied France, gasoline was a scare commodity, and Gagnan had been working on a device that would allow cars to run on compressed cooking gas. His "demand regulator" automatically adjusted the pressure of the gas as it flowed from the tank into the motor. The same principle, Gagnan said, could be applied to the pressurized air entering a pair of human lungs underwater.

Gagnan built a modified version of his regulator in only a few weeks and stood on the bank of the Marne River while Cousteau field-tested it. The science of metallurgy had come a long way since the days of the *aerophore,* and the metal in Cousteau's and Gagnan's air tank could withstand greatly increased pressures and thus could contain far more air. Wearing this tank strapped to his back, with Gagnan's regulator attached to its valve, Cousteau waded expectantly into the Marne, breathing from the regulator mouthpiece as the water closed over his head. The device worked, but imperfectly. Only when Cousteau swam horizontally was there an even flow of air. When he was upright, air streamed out of the exhaust pipe, and when he was upside down he could hardly breathe. On the drive back to Paris the two inventors brainstormed the problem, until in a flash of inspiration they realized that the flow of air was erratic because the intake and exhaust valves of the regulator were not on the same level. According to Pierre de Latil, they actually said "Eureka!"

Gagnan went back to the workshop and made an improved version, which he sent to Cousteau by rail. "No children," Cousteau later wrote, "ever opened a Christmas present with more excitement than ours, when we unpacked the first 'aqualung.' "

Inside the box were three diving cylinders, bundled together and attached to the improved version of the regulator, which sprouted two floppy hoses that connected at a mouthpiece. One hose brought air to the diver, the other flushed his exhalations into the sea. With Dumas, Tailliez, and Cousteau's wife, Simone, watching, Cousteau hoisted the three cylinders onto his back. "Staggering under the fifty-pound apparatus," he remembered, "I walked with a Charlie Chaplin waddle into the sea."

Cousteau reached the bottom "in a state of transport." The aqualung worked. He swam into coral tunnels, turned somersaults and loops in the water, all the time receiving an even flow of air. "From this day forward," he wrote rapturously in *The Silent World,* "we would swim across miles of country no man had known, free and level, with our flesh feeling what the fish scales know."

All of them tried it. Simone astonished a fisherman by emerging from the sea with a lobster in each hand and asking him to watch them for her while she went back for more. "Back on shore," remembered Tailliez, "we danced for joy." Later Tailliez even made an aqualung for his dog.

The diving equipment I used at Grand Turk was no different in principle than that invented by Cousteau and Gagnan almost fifty years earlier. Sometime in the 1960s the old floppy double-hose regulators gave way to a refined single-hose version, and the necessity for double and triple air cylinders faded with the introduction of higher-volume air tanks, like the yellow aluminum one I was wearing. Fashions in masks and fins came and went, dive knives evolved from gruesome saw-toothed implements to elegant utensils no more threatening than fingernail clippers, spear guns were thankfully almost extinct, and buoyancy compensators now helped divers retain neutral buoyancy in the varying water density. But in all important respects the technology for recreational diving has remained at a standstill. "The lung is primitive," Cousteau wrote of his creation at the end of *The Silent World,* "and unworthy of contemporary levels of science."

As much as I loved the Aqua-Lung, I had to agree with its inventor. I found it a burdensome and limited apparatus. Coasting along at sixty feet, I could still feel the unwelcome weight of the tank, and occasionally its valve would bump up against the back of my neck. The noise and fuss of my own exhalations annoyed me — all those bubbles constantly announcing my presence. From time to time a bit of water seeped into the bottom of my mask, and I had to clear it by holding it tight against my face, cracking open the bottom seal, and blowing through my nose. Despite my best efforts to keep them out of the way, various hoses floated into my line of sight or became tangled with my arms. And then there was the aggravation of knowing that all this equipment could take me only so deep and allow me but a short stay.

The problem, as always, was pressure. If I went too deep breathing compressed air, the nitrogen in my tank would slowly poison me and I would succumb to rapture of the deep. And if I survived nitrogen narcosis, I would still be at risk for the bends, also known as caisson disease or decompression sickness. This deadly condition is as old as human exploration of the sea. It was first noticed in the laboratory in 1670 by Robert Boyle, the English scientist whose great and lyrically titled book *New Experiments, physico-mechanical, touching the spring of air, and its effects,* introduced the principle known as Boyle's Law: that the volume of a gas is inversely proportional to its pressure. Some of Boyle's experiments involved observing animals in a chamber of compressed air; once, as he lowered the pressure in the chamber, he observed a peculiar gleam in the eye of a snake. The gleam was an air bubble, and though Boyle did not know what caused it, we now understand that the snake was, in diver's parlance, "bent."

It took another two hundred years to fully understand why helmet divers and workers who labored on bridge pilings and tunnels in compressed-air chambers called caissons kept falling victim to a mysterious ailment when they returned to the surface. Many of them died, others were crippled for life. In its mildest form, the condition manifested itself as an intolerable itching sensation, but in more severe cases the victims suffered lacerating pain in their joints that drove them into contorted postures — "the bends" — in a vain effort to relieve the agony.

Paul Bert, a French physician, plastic surgeon, and champion of women's suffrage, was intrigued by the problem. He interviewed victims of the bends and experimented on himself, his wife, and various laboratory animals in an air pressure chamber. In the end the culprit he fingered was nitrogen. Not only does this gas — which makes up four-fifths of the air we breathe — become inebriating when breathed at a certain depth, it also saturates the body underwater. The deeper a diver goes, the more nitrogen builds up in his tissues and blood. This is no problem until he starts to come up, when the decreasing water pressure causes the gas to come out of solution. If he ascends slowly enough, the nitrogen is reabsorbed without further complications. If not, it forms bubbles that fizz the blood and block circulation.

For the sport diver, the way to avoid this predicament is simple: you don't go too deep, and you don't stay too long. You allow enough time between dives for the excess nitrogen in your blood to bleed itself out. On a deep dive, it's often necessary to plan decompression stops on the way back up, releasing the nitrogen as it comes out of solution, before the diminishing pressure has the chance to turn your blood into a Coke float.

I had not gone deeper than eighty feet today, and technically did not need to worry about the bends, but I took my customary ten-minute decompression stop at ten feet on the way up. Looking down at the reef, I saw the bubbles of two other divers in our group, hotshots with four-hundred-dollar decompression computers that would beep when it was time for them to head to the surface. Those computers represented the momentary cutting edge of diving technology, but I yearned for something more radical, some breakthrough that would take me beyond the barriers of pressure and air supply. I wanted to be a true manfish, to dive down with no awkward tank on my back and no limits to my time. It was demoralizing to have to come up after only fifty minutes or an hour, to unstrap all my equipment and to sit around in the boat waiting for all that nitrogen to waft out of my pores before I could go down again.

Underwater Man is on the drawing boards. In laboratory experiments, unfortunate rats have been made to breathe a superoxygenated liquid with a precise saline content, though that is a long step away from having them somehow gulp air from normal seawater. Machines have been envisioned that would effectively act as a lung bypass, allowing a diver to breathe underwater by pumping his blood through a device worn on his belt like a Sony Walkman. Cousteau longs to behold "the ultimate hero of *Homo sapiens,* a naked man mingling long and deep among the fishes, breathing water as they do," and in his writings he indulges in some speculation as to how such a creature might be cobbled together:

> It would be necessary to surgically insert semipermanent injection tubes in the windpipe; provide heated diving suits; fill the lungs with an oxygenated, carbon-dioxide-absorbing liquid while the diver is under anesthesia; connect auxiliary tanks and circulatory pumps to the injection tubes in the windpipe; fill internal cavities with a saline solution; and maintain a standby team of medical assistants. Then such a person would be capable of diving for as long as six hours and as deep as 3000 feet.

Any volunteers? When I first read the description of this grotesque operation I shuddered with revulsion — but also with excitement. Cousteau, for his part, dismisses the inconvenience. "To backbreed man," he writes elsewhere, "and restore his gills may not be as terrible as the supreme crisis of his life struggle, the moment he is born."

My ten minutes of decompression were up, my nitrogen was venting nicely, and now it was time to laze upward to the surface. As I ascended, following the disruptive stream of my bubbles, I wondered what I would really give, what I would really endure, to become a manfish.

14

Homesick Turtles

I went to see an old man named Oliver Lightbourne, a turtler and conch fisherman from the old days. He lived in a ramshackle compound on the hill above Sambo's Chicken, at the south end of North Creek. When he came to the compound gate he was wearing a Hawaiian shirt and ancient red double-knit pants sliced off just above the ankles. He was eighty-one, wiry and wizened, with a small black dome of a head.

"I got some people here now," he told me, not inviting me in. "I don't want them to know my business. You come back tomorrow, two o'clock, with the help of God, our heavenly father. You catch my meanin'? If we both be alive, you come back."

I got back on my scooter and drove over to Mitch's in time to help load the tanks for the afternoon dive. Swimming through the canyons of the wall that afternoon, I had sea turtles on my mind. The day before, I had startled a foot-long adolescent, a hawksbill or a green, as it snoozed beneath a coral rock. It rocketed away too fast for me to be able to determine its species, but I was entranced by the almost supernatural grace of its flight through the water. It was as if a heavy stone had suddenly sprouted wings and soared away.

Now, as I drifted through a long tunnel, I glimpsed another turtle sailing through the bright blue window ahead. This one was larger, three feet long, and it swam with the flitting motions of a swallow. When I emerged from the tunnel I hung there and watched it. Like the turtle I had seen the day before, it was either a green or a hawksbill; I wasn't enough of a naturalist to identify it on the fly. But I knew that a green of that size would probably be grazing in the turtle grass beds, not stroking through the deep-water coral formations. Hawksbills, on the other hand, prefer coral environments, where they feed on sponges.

So this was most likely a hawksbill, I thought as I watched it glide in and out of the underwater scenery. The sight of it filled my heart. Like a flock of snow geese gliding overhead on a spring evening, or a mountain meadow brilliant with wildflowers, the turtle struck me at that moment as one of the defining beauties of nature. It kept its distance — it must have been forty or fifty feet away from me — and I did not try to follow. But I could see the turtle's long front flippers stroking evenly, pulling forward while the rear pair stayed tucked beneath its shell. I could see its torpid, uncomplaining expression. Even so, the distant water shrouded it and made it seem hazy and fleeting, like something you might catch sight of in a dream and urgently, even desperately pursue without knowing why. Because the turtle's flight moved me so greatly, because it seemed so gloriously unreal, I checked my pressure gauge, thinking I must have drifted deeper than I had realized and was suffering the effects of nitrogen narcosis. But I was only at sixty feet.

The next day, at two o'clock, I went to see Oliver Lightbourne again. This time he welcomed me into his compound, and we sat in a little porch area built out of plywood and corrugated metal. Off the porch ran a maze of shacks, which Lightbourne rented out to Haitians. He himself had been here, he said, since 1926, and the house that formed the basis of his compound was originally a relief house from the government. Besides being a landlord, he repaired radios and watches. He took a seat behind a table filled with old transistors and the husks of table radios left over from the forties and fifties. Through the slatted fence of the compound I could see a few rows of scraggly corn and a sea grape tree. In a folding chair sat a middle-aged woman, smiling slyly at the idea of a visitor; next to her a fourteen-year-old girl in a school uniform and glasses ate Pringles from a tube.

"So," Lightbourne said, "you want to know where be the best place to find conch."

Though I hadn't asked that, I nodded anyway.

"Of course I can tell you this. The east part of the lighthouse have a big reef. They got a grass on the south side of that reef name the Basin. After you leave from that place you go east, a place by the name Bay of Mexico. It got a very very shallow water there. The south side of that big grass be named Jack Shoal. After that you meet a shoal name Long Bar. You go east from there you meet a place name Steak Shoal, then to another place called Cox's Reef, then to another sandbar. Then to the Blue Hole. Whens you get there you go south to the creek. That's where there be conch. Oh, there was lots of conchs there, once long ago.

"Diving for conch, we use only the bare eyes. Not the

goggles. I could bring up six conch at a time. Two in this hand lip to lip and four along my arm. I was strong like that, once long ago. I take a deep breath and after I feel the breath com short I com out the water. I keep a good breath. Under water I con blow out breath, but I con't suck in, cause then I be suckin' in salt water. You cotch my statement?"

He sat back in his chair and closed his eyes. "They had a whaleboat here once," he said, "but that's long years." Then he began to ramble on about his life as he tinkered with the radio on the table. His father, he said, was a merchant seaman who had sailed around the Horn on a voyage that took two years. Later he moved his family to New York, where Lightbourne said he grew "from child to man." "I went to P.S. 68, between 127th and 128th. You know that area of New York? I used to could call all the streets in that city."

He said that during World War II he worked on a ship carrying dynamite and TNT from Brazil to New York. "Lots of ships got torpedoed, but I got clear. They sink two in front of us. When I was on that ship with that dynamite I was thinkin' I would see this land no more. I don't know how I got clear, but God's the boss."

He listed the places he had been in his life: the Amazon, Curaçao, Cartagena, Colón, Maracaibo, Trinidad . . . Then he grew silent, contemplating the antique radio parts. I asked him about turtles, and he snapped back to attention.

"For tortle," he said, "you get to Cox's Reef. They have a big grass there. Shoals all around. Grass be growin' dark on the bottom. Dark. Dark. Place called Goggle Eyes.

"To cotch the tortle, you have a net you hang like a curtain. You put it out in daylight. Sometimes the sharks go in it. Sometimes the sharks eat the tortle. When you see the tortle in the net and he raises his head and sees you he goes down. You wait til he com up again, then you grob him by the fins."

Lightbourne rummaged around in a pile of radio debris and held up the shell of a green turtle. "If you grob him this way," he said, holding the shell by its sharp sides, "your whole skin gone. You got to stay away from his mouth too. If he bites you, you lose that piece."

Lightbourne still caught the occasional green turtle. He sold the meat to neighbors and usually was able to sell the shells for fifteen or twenty dollars.

"Hawksbill shell," he said, "is more valuable. The flakes more different and pretty. Once long ago, the flakes were cheaper. The meat was cheaper too. Everything was cheaper in this land, but more people increase here and by and by the price go up."

I stared at the shell that Oliver Lightbourne held in his hands. It was dusty, and the turtle's spine still clung to its underside. I suppressed a hideous image of that creature lying helpless on its back as its throat was sawed open with a knife and its plastron cut away from its body, revealing a woozy mass of viscera. There was no thought more gruesome to my imagination than that of a turtle being pried from its shell. I'd read that in some parts of the world live hawksbill turtles are placed upside down on coals so that the searing heat will make it easier to remove the upper layer of their shells. The naked turtles are then thrown back into the water in the belief that the shells will regenerate. They do not, of course. The mutilated turtles die a slow, savage death. It is easy to understand, however, how this belief could arise, because sea turtles have an aura of dogged immortality. They can live to a hundred years, perhaps much longer, and they do not surrender all that life easily. As those who prey on them have long noted, turtles die hard. They seem to care more than their tiny brains should allow them to. They gasp in agony and shock, they snap furiously with their jaws, and after they are dead their flippers still wave pathetically, stroking through the absent sea.

All the world's species of sea turtles — greens, hawksbills, loggerheads, ridleys, and leatherbacks — are in serious decline, some of them teetering on the edge of extinction. It is illegal to import turtle products into the United States, but when I was in Grand Turk you could still order turtle steak in a restaurant. I never did, though a dinner companion once goaded me into taking a bite of his. Green turtle meat is supposed to be one of the greatest delicacies on earth, but my guilt prevented me from even noticing the taste. On the subject of sea turtles I was prim and high-minded. These lugubrious creatures weighed on my conscience and excited my interest in ways I couldn't explain. They had some kind of magical charge for me, and I understood the impulse that drove certain Mayan artisans to create stone turtles whose heads were carved from magnetic rock. Sea turtles were living, wandering lodestones.

When Columbus came to this part of the world, sea turtles were abundant. He observed Indians capturing turtles by tying a leash to a remora, a fish that attaches itself to sharks and other large ocean-going animals by means of a suction pad on its head. When the remora adhered to a turtle, the Indians would pull it in and later toss it pieces of turtle meat as a reward. On Columbus's last voyage in 1503, he encountered the Cayman Islands, which he named Las Tortugas because of the thousands of green turtles that crawled ashore in the summer to lay their eggs. The turtles made the islands an important port of call for generations of voyagers, but the Caymans were not the only places they were found. Sea turtles thrived all over the Caribbean, the Gulf of Mexico, everywhere. To the Spanish, French, British, and Dutch sailors who colonized these islands, the creatures were blessedly easy prey. On the beach the heavy turtles were helpless. They could be flipped over and then hauled at leisure to the ships, where they were stored as living provender, kept on their backs without food or water for weeks or months.

Slowly over the centuries, as more and more turtles were taken for meat, for soup, for jewelry and leather, as their eggs were dug up by wild pigs and dogs and by humans searching for aphrodisiacs, the sea turtles of the world began to disappear. The last glimpse we have of their former population strength is a grainy home movie from 1947, filmed at a Mexican beach known as Rancho Nuevo. The film shows the last great *arribada* — the "arrival" of perhaps forty thousand Kemp's ridley sea turtles upon their ancestral nesting beach. The turtles have never reappeared on that beach — or another — in anywhere near those numbers. (No more than four hundred nesting females now make their way to Rancho Nuevo, for instance).

As many as four million baby turtles might have hatched from that single *arribada*. Female sea turtles tend to nest in cycles, swimming in the ocean for one to four years and then coming ashore as many as five times in one season to lay perhaps a hundred eggs at a time. It is believed that the turtle returns to the beach of her own birth, swimming sometimes a thousand miles to reach it and crawl up onto the sand only yards from where she deposited her last nest. The males, in all their long lives, never come ashore at all. But during egg-laying season they congregate in the waters off the beach, growly with lust. "Sea turtles in love," writes the zoologist Archie Carr, "are appallingly industrious. . . . To hold himself in the mating position on top of the smooth, wet, wave-tossed shell of the female, the male employs a three-point grappling rig, consisting of his long, thick, recurved, horn-tipped tail, and a heavy, hooked claw on each front flipper."

Some species of sea turtle — loggerheads, for instance — mate with more serenity than others, but often the short-sighted males grow crazy with desire. Leatherbacks, in particular, will sometimes try to mate with anything bearing even a superficial resemblance to a female turtle.

"SEX-CRAZED TURTLE TRIES TO MATE WITH SCUBA DIVERS" screams the headline of an article I found in a tabloid newspaper several years ago. "The damn thing tried to rape me," a real estate agent in Islamorada, Florida, is quoted as saying. According to the newspaper, the man got away from the "giant gay turtle," but another diver, who "wished to remain anonymous," wasn't so lucky. He was "mounted" for five or ten minutes.

I have no idea whether this is true, but it's the kind of thing you hear about blundering male sea turtles. The females, though, are even more single-minded. Though they are skittish when they first haul themselves up on the beach, once they have dug their nests in the sand and have begun to lay eggs they cannot be distracted. Photographers may set off strobes in their eyes, grinning yahoos may jump up and down on their backs, wild dogs may eat the eggs as soon as they come out, but the turtle pays no attention. Typically she scoops out a shallow hole in the sand with her front flippers, and when she is settled into it she carefully sculpts a deep egg chamber with her hind flippers. She then lays her eggs, depositing them through an opening whose wonderfully apt name is the ovipositor. When she is through she fills the nest in, scatters sand about to camouflage it, and tamps down the roof with pounding motions of her body. Then she turns and heads back to the ocean, a groping, wheezing dinosaur, with all her great weight — three hundred, four hundred, a thousand pounds — pressing her down into the sand. As she progresses, she gouges out a trail, and saline tears stream from her eyes. Her awesome duty done, the mother sea turtle never looks back.

The eggs are soft and leathery; depending on the species, they range in size from Ping-Pong to tennis balls. There is hardly a creature in the vicinity of the beach that does not consider them a delicacy. Raccoons, ghost crabs, dogs, coatimundis, coyotes, peccaries, runaway cats, and various sea birds all prey on the eggs — and, on a massive scale, so do humans. One reason the great *arribadas* of Rancho Nuevo no longer occur is because the nests were wiped out by local inhabitants. Families would pick a turtle and shadow her, patiently waiting for her to lay her eggs so they could steal them. Egg entrepreneurs flocked to the beaches, dug up the nests, and took the eggs away by the millions in pickup trucks and mule trains. On the Florida coast, turtle eggs — whose whites do not coagulate when cooked — were a staple ingredient in waffles and pound cakes, and local bakeries had their own egg-harvesting crews that drove down the beach looking for "crawls" — the broad tracks made by gravid sea turtles hauling themselves through the sand. The eggs that were not used for baking were eaten as aphrodisiacs. They were displayed in towering pyramids in Mexican bars and sold in Florida saloons for as much as a dollar an egg to men in need of a sexual jump-start. "Every Saturday night," Jack Rudloe writes in *The Time of the Turtle*, "the streets and gutters in front of these sleazy saloons were littered with piles of turtle eggshells."

The eggs were eaten raw with a consoling chaser of beer. The taste was so powerful, so intriguingly foul, that merely surviving the experience of eating them brought people illusions of potency. Rudloe, in an appealingly bizarre passage in his book, describes how he ate an egg straight out of a nesting turtle's ovipositor. "God, it was strong!" he writes, but then, "I felt new blood surging through my veins. I felt the wind blowing in my face. I heard the sea roaring out there. I felt great."

If their eggs are not discovered, the baby sea turtles hatch several months after their mother has crawled back into the ocean. If you were to put your ear to the sand at such a time, you could hear them rustling down below, clambering upon each other's backs in a blind attempt to reach the top of the nest. All this activity dislodges the sand in the chamber, which drifts down to raise the floor of the nest higher and higher. Finally all the hatchlings — a hundred baby turtles — break through at once and stream out of the sand, their flippers batting the air. The baby turtles move with comical gracelessness. They may look at first as if they have no idea where to go, but almost invariably they flap their flippers and orient themselves toward the ocean. Nobody knows precisely what leads them there, though the stimulus is apparently a visual one. Something about the light above the ocean, or the light within it, stirs them and draws them forward. If there is phosphorescence in the waves, if the whitecaps glow in the moonlight, they quicken their already frantic pace. They need that gleaming ocean the way baby mammals need milk — it's all they know to search for. They sense that if they can reach it, they're home.

But many of them die on the beach, where ghost crabs drag them into their burrows and pick out their eyes, where coyotes crunch their shells in their teeth and various seabirds pluck their wriggling bodies off the sand. Once the hatchlings fling themselves into the waves they are not much safer, since a host of carnivorous fish waits for them just offshore. Of the hundred hatchlings, perhaps none will survive, perhaps one or two. For their first several years they will drift along in the ocean currents, too small to plot their own course against the flow, too vulnerable to stray far from the floating thickets of seaweed in which they can hide and search for food.

In 1981 I participated briefly in a "head start" experiment for ridley turtles. The idea was to establish a nesting beach on the Texas coast to augment the precarious rookery at Rancho Nuevo, which could be wiped out by a single hurricane or oil spill. Because sea turtles obstinately return to the beach where they were born to lay their eggs, it was thought that eggs could be taken from Rancho Nuevo and incubated at Padre Island National Seashore in Texas. Once hatched on Padre Island, the baby turtles would be imprinted with the memory of a different beach, to which they would return eight or ten years hence when they were ready to lay their own eggs. All of this was theoretical. No one knew if imprinting was really the mechanism that caused turtles to return to their natal beach, and no one was certain how long it took a turtle to reach sexual maturity — it might indeed be eight or ten years, or it might be twenty, or even fifty. Nevertheless, workers at Rancho Nuevo sidled up to the laying turtles and collected their eggs in plastic bags, then buried them in Padre Island sand and delivered them to their new natal beach. When the turtles hatched, their attendants followed them to the water, allowed them to develop a feel for swimming, and then scooped them up in nets and sent them to Galveston, where they were reared in buckets until they were the size of dinner plates and their vulnerable first year was behind them.

At the end of that year 1,520 yearling turtles were ready to be released into the Gulf. The turtles were marked, placed in cardboard boxes at Galveston — eight to a box — and delivered to Port Aransas via Coast Guard transport plane and U-Haul truck. The boxes were then loaded onto a seventy-two-foot shrimp boat. I went along on the shrimp boat to help throw the turtles overboard. Awaiting release, the turtles poked their heads through the slots in their boxes, looking straight ahead with their unblinking eyes. We opened the boxes, picked them up, and tossed them like Frisbees into the Gulf. Each turtle had the same reaction to being thrown into the ocean. It would wave its flippers frantically in midair and then, when it hit the water with a belly flop and began to sink, it would heave itself up to the surface. The turtle would stay motionless there for a second or two until it had gathered confidence and then head out into the open ocean with secure, efficient strokes.

I learned years later that some of the turtles I had thrown overboard in the Gulf of Mexico that day were captured in the Mediterranean, but after ten years none of them had returned to Padre Island to nest. Sometimes, though, I have seen the carcasses of sea turtles washed up on the beach at Padre, their shells bleaching in the sun and sand crabs crawling through the eye sockets of their skulls.

What happens in a turtle's life between the time it enters the water as a hatchling and its solitary death is largely a matter of speculation. All sea turtles, apparently, are wanderers, paddling thousands of miles through the empty oceans, obeying the mysterious agenda inside their simple brain. Whenever I saw turtles swimming in the waters of Grand Turk I felt a motherly tenderness toward them — for their unexpected grace and their dim-witted, persevering mien. The range and magnitude of their lives — all those years wandering through the ocean — struck my human imagination as an unbearable burden, a crushing load of time and tedium. And their implacable faces left me with the impression they were on some sort of hopeless, broken-hearted quest.

In some parts of Mexico, when the turtles crawl ashore to lay their eggs, the event is known not as an *arribada* but as a *morrina,* a word that means, roughly, "homesick." If sea turtles have one overriding compulsion, it is to get home, back to the beach where they were hatched, to the specific stretch of sand where they will lay their own eggs and copulate offshore. The intensity of this drive is beyond our comprehension, as is the turtle's ability to find its way through so much trackless ocean to a nameless fragment of shoreline or a barely discernible midsea islet. There are many hypotheses about how they navigate. Possibly they steer by the sun, or even by the stars. They may smell their way home, detecting the chemical essence of their destination as it drifts in the currents. Their brains may be receptive to the magnetic fields given off by certain underwater landmarks. In all probability they use a combination of methods, moving at first in a general direction, following the moving compass of the sun or the sweep of a current, then fine-tuning their travels by rising to the surface and somehow reading in the wave patterns the anomalous presence of a distant island.

Ridleys are the smallest of the sea turtles, and the swiftest. It is unusual for them to reach a hundred pounds. The largest by far are the leatherbacks, which grow to eight feet or more in length and can weigh a thousand pounds. Their shell is soft, ridged, and tight fitting. Leatherbacks are loaded with fat to protect them during their peregrinations into subarctic waters. They eat mostly jellyfish and have backward-facing spines in their throat to keep their slippery prey from working its way back up. Of all the sea turtles, the leatherback is the only one that does not have staggeringly bad breath. In these days of widespread ocean dumping, their diet puts them at grave risk, since they are likely to die from intestinal obstruction if they mistake a floating plastic bag or the clear plastic rings from a six-pack of beer for a form of gelatinous sea life. When pestered by divers the leatherback has proven to be as short-tempered as its prehistoric appearance would suggest, chomping and flailing like a colossal snapping turtle.

Almost every day for a week I saw one or two turtles coasting along in the underwater distance, buoyant and agile. I could not get over the beauty of their swimming, the smooth passage of those fluted shells through the water. The turtles kept their distance and were usually too far away for me to identify their species. Frustrated, I daydreamed about the possibility of inventing an underwater pair of binoculars. With such an instrument I could drift down to the sand beds and watch the flying turtles overhead the way a birder studies distant hawks as they spiral above a ridge line.

Once a big hawksbill crossed my field of vision ten or twelve feet away, close enough for me to take flight and swim along with it. The turtle was in no hurry, but I had to stroke hard with both fins in a dolphin kick just to keep from falling hopelessly behind. I felt the awkward weight of the tank against my arched back and marveled at the turtle's shell, which despite its size seemed to weigh nothing. Far from being a burden, like my air tank, that the turtle was doomed to drag through the water, the shell seemed in some way to be the source of the creature's ethereal mobility. Tracking the turtle, I could see its lozenge-shaped eyes gazing indifferently ahead. I wondered what signals it was following — whether it was reading the chemical content of the water or the position of the sun or the magnetic pull of the earth. It did not have much of a brain in its head, but it did have a map. The turtle could find its way through a thousand miles of open sea, relying on signposts and signals that a human could never understand, driven by a homing instinct so powerful that my own ruling compulsions and desires began to seem petty in comparison. The turtle was a great being, venerable, unknowable.

Once, diving at forty feet, I came across what I assumed was an algae-covered boulder about five feet long. But the boulder was moving, rocking perceptibly with the surge, so I knew it had to be a turtle. Was it dead, and was this its empty shell lying on the bottom? I could not see its flippers or its head, but I noticed that the front end of the shell was wedged beneath a coral ledge. Gently I poked my head under the ledge. In the cavelike darkness, I could see, inches from my face, the mighty cranium of a loggerhead turtle. The turtle was asleep; its eyes were closed. I held my breath in wonder, then exhaled slowly and carefully, not wanting my bubbles to wake the loggerhead. Possibly this was dangerous. If I surprised the turtle, it might become frightened and disoriented and lash out at me with its jaws — jaws strong enough to break through a conch's shell. But it seemed deeply asleep, frozen in prehistoric slumber. I watched it for a long moment more, and then I floated away, trailing my fingers lightly across its shell.

15

Wandering

One afternoon I rode the scooter south past the airport and the old Pan Am buildings and took a dirt track that led seaward through a shallow limestone canyon. When the track ran out I grabbed my bag of gear and walked the rest of the way to the shore in my flip-flops, hoping to find a new and interesting place to snorkel. The gray rock had been honed to sharp pinnacles by the action of the water, and it was hard to find a safe place to put my feet down. The rock underfoot seemed brittle and hollow, undercut by waves until it was nothing but a thick crust.

There was no beach, just foamy water slamming into jagged rock. Not a prime snorkeling spot, but the view was restful. I could see the waves breaking on the sickle-shaped beach at Gibbs' Cay a half mile away, and between the shore and the cay a lone fisherman was standing up in a small skiff, poling it along through the shallows. I sat down in the shade of some brush, with my gear bag for a pillow, and drank a bottle of Gatorade. Then I let myself sink into that half-conscious hallucinatory state that precedes real sleep. I could feel my body rocking in the surge as a parade of fish passed in front of my closed eyes. They swam up one by one — parrotfish, wrasses, hinds, and groupers — and peered at me without curiosity. A queen triggerfish stared the longest. The yellow bands of color on its head looked sinister and deliberately applied, like a cannibal's tattoos. A conch swam by, not crawling on the bottom but drifting through the water with the buoyancy of a nautilus, its eye stalks waving. And in the cloudy blue distance was a human figure, clad head to foot in neoprene, with Leonardo da Vinci frog gloves on his hands. He hung motionlessly in the surge as if dead, but the ocean was filled with the roar of his breathing.

By the time I snapped awake from this shallow dream, the sun had moved a few degrees lower in the sky, robbing me of my patchy shade. I could feel its warmth weighing on my eyelids and tightening the skin over my cheekbones. It was time to get moving, but the scooter wouldn't start. It was out of gas.

I pushed it down the dirt track, through the limestone canyon, and out onto the road fronting the southernmost salina. There was no shade anywhere, and sweat cascaded through the headband of my baseball cap and down into my eyes. Wiping my stinging eyes with my shirt, I realized there was more salt in my own sweat than there was in the seawater in which I had been diving. I was angry at myself for running out of gas, and as I forced the scooter down the soft clay road I began to lose patience with the compulsion that had brought me here. I had been underwater for so long that my perspective was warped. I noticed, when I looked at the gnarly, flattened tops of the island brush, that I automatically registered it as elkhorn coral. I was so used to looking at fish that it took my brain a second to process the cattle egrets and black-necked stilts along the roadside as birds. In a way, I thought, I had what I wanted: I was living underwater. My imagination was soaked in undersea thoughts and imagery, and here on the surface world it operated with a perceptible lag — refactoring, reconfiguring, purging itself of an endless store of aquatic references.

After a mile and a half of dragging the scooter along the roads I came to the Kittina Hotel. I went inside and ordered two Cokes. The open-air patio was shady and calm. Today's lunch special — oxtail soup — was still written on the fluorescent blackboard, but it was almost two o'clock by now and all the diners had left. Only John Houseman, former editor of the *Conch News,* was still around.

"You're an educated man," he said, slipping into the chair across from me. "What is the difference between 'pulling the forelock' and 'knuckling the forehead'?"

"I have no idea."

"Well, of course you wouldn't know," he said, "it's *very* European."

Houseman was seventy, or somewhere close to it, a thin man with a bit too much twinkle in his eye. He was wearing a threadbare synthetic shirt and glasses that hung from his neck by a string. He had a disheveled, twilight-of-the-empire look, with a swept-back mane of gray hair and a giant mustache.

I asked him how he had come to live in Grand Turk.

"Well, of course I was born in London, you see. Went to the right sort of schools. *Of course* I went to Sandhurst, *of course* I went to Oxford, *of course* I joined the right regiment. Became an archeologist in Greece! Then of course when the Colonels took over I made the transition — came over here to the Turks, started my paper. Labor of love and all that. Lost about sixty thousand dollars. But at one time we had a print run of seventeen hundred and fifty — two-thirds of that going overseas! And of course I did a radio broadcast — all about toilets and shit houses and things like that. I think I was perfectly right to talk about it, don't you?"

"No question," I said.

He lifted his glasses from his chest and examined their wobbly frame. "The little screw's rusted out. Metal things fall to pieces here. May very well be that we'll have to go back to sisal. Four thousand miles of trade winds! Nothing between us and Africa. Nothing to absorb the salt, you see.

"But I'm very happy here. People know me and I know them. People don't come here to see *things*. We have no *mausoleums*. The only thing we have here is a very strange collection of people. It's an ugly place, it's not a pretty place, but it's got some interesting people. Well, like you — you're a bit of spice in the pudding. But we don't take you too seriously, do we?"

His eyes flashed open. "I'm suddenly reminded! I must go and have a pee!"

When I went diving later that afternoon the water was clear and sun-struck. Swimming along on my back at forty feet, I could see a frigate bird in the sky above, distorted by the mild surface swells into a Cubist apparition. I glided down to Harmonium Point with my mood restored. Harmonium Point was as familiar to me now as a childhood landmark, a welcoming place filled with benevolent secret powers. I swam along through the canyon next to it, checking on the secretary blenny and inspecting the fiery red carpet of encrusting sponges, which grew on the rock like a fungus and were painful to the touch.

Sponges were everywhere on the reef — not just spreading on the coral rock like the encrusting sponges but sprouting outward in waving fingers or tubes or vases. One of the basket sponges near Harmonium Point was big enough for me to put my head into. Its interior walls, though hard to the touch, appeared to have been spun from some insubstantial filament. It was a deep, musty shade of gray, with the rich texture of objects seen through an electron microscope.

Sponges are animals, hard as it is to imagine. They have no organs and no skeleton, only a kind of hardened infrastructure made up of Tinkertoy units known as spicules. The sponge exists by taking in water through holes in its skin, then channeling it through interior passageways lined with specialized cells that scrub it free of plankton and send it out again. That's it. Not much else goes on in a sponge, except for some low-level reproduction. But this simple creature is one of the first things that drew divers down into the sea. In the ancient world thousands of sponge divers harvested this all-purpose utilitarian organism. Sponges were used for more than washing and scrubbing. Soldiers used them as helmet liners and soaked them in water as canteens. Mothers gave honey-soaked sponges to their children as pacifiers.

When I looked inside one of the big basket or tube sponges I almost always saw a host of other organisms — brittle stars, small crabs or shrimp, and territorial fish. And when I put my face close enough to its skin I could feel the waste water rippling out of the sponge like the exhaust from a jet engine.

Most of the plankton that it processed were invisible to me. Sometimes plankton made the water of the reef cloudy, but today the water was so utterly transparent it seemed like a sparkling vacuum. Even so, it was soupy with life. The plankton constituted an unseen world, an endless universe filled with a bewildering creation of animals and plants that to my eyes was hardly there. I remember once, at a marine lab in Florida, looking at a single drop of seawater under a microscope. Half a dozen gossamer crustaceans — copepods — were lurching about inside it, swimming this way and that at sharp angles, their feelers waving ahead of them and their legs bristling with motion. When they surged downward, into the depths of that single water drop, I could see them grow smaller. Compared to the sea in which they swam, they were so tiny that they could actually sink out of my sight. But when they came back to the surface of the drop I could make out their black eyes, and I watched them hunt down the snowflake bits of one-celled plant life known as diatoms. Eventually, after four or five minutes, I noticed that the copepods were not quite so active; their whipsawing feelers had slowed to a lazy sweep. Then they were dead. When I lifted my eyes from the microscope I felt odd, and absurdly contrite.

From our perspective, copepods are just grainy bits of life, but to much of the rest of the plankton world they are lumbering beasts. Most of these creatures are far more minute and far simpler. There are two kinds of plankton: phytoplankton, which are plants, and zooplankton, which are animals. Diatoms are phytoplankton, as are dinoflagellates, one-celled plants that beat through the water by means of a whiplike tail. Individually, these plants are microscopic, but sometimes they flourish in such sudden abundance that the surface of the sea is discolored by their presence. During the daylight hours phytoplankton rise upward in the water column in order to capture sunlight for photosynthesis.

Zooplankton graze on phytoplankton the way cows graze on grass. They also prey on each other. A planktonic animal can be almost anything: a single-celled protozoan, a larval octopus, a crustacean, a coral planula, a jellyfish. What binds these creatures together in our minds is the fact that their lives are controlled by the currents. The word plankton means wanderer. We think of them as helpless, passive microdots being swept along by forces they cannot control, delivered to destinations they have not chosen. In fact, they all have some control of their movement. They prowl and hunt, moving along by spermlike tails or oscillating hairs or propulsive contractions of their billowy tissues. They move on a scale that is congruent with their lives, as uncomprehending of the vast ocean currents that rule their destinies as we are of our own universal matrix of space and time.

Salt water began to leak into my mask, and I took it off at sixty feet as casually as I would take off my glasses to clean them on a napkin. Tightening the straps, I looked around at the fuzzy blue fabric of the reef. Without my mask, it was like an extremely snowy picture on a color television set. Shapes were vague, colors were smudged, and there was no perception of depth, just a flat, nappy backdrop. But when I put the mask back on and cleared out the water, the world instantly snapped into a focus so sharp I felt a euphoric rush. Near my elbow a blue tang was nibbling on a stalk of coral as if it were an ear of corn. A moon jelly floated by. Beneath its diaphanous mantle, I could see the elegant rhomboid shape — sleek as a corporate logo — of its interior structure. A butterflyfish, its long snout tapering like a pair of forceps (to aid in extracting organisms from coral fissures) swam across my line of sight. It had a big parasitic copepod clamped on each side of its face. The copepods looked like pillbugs — they were black and sheathed in a reticulated carapace — and I turned away from the fish as if from some grotesque affliction in a fellow human.

While the other divers were surfacing and waiting their turn to climb back onto the boat, I made a random sweep of the bottom. I was at home again, here alone on the reef, with the other divers up there on the surface floating like helpless balloons. I felt smug, strong, secure: the manfish, the water baby. Nothing in my life had I mastered so completely as this, the strange ability to be comfortable underwater and to call every creature I saw by its name.

Here, for instance, was a sea cucumber, an echinoderm related to the starfish or the sand dollar. It was a drab but suggestive-looking thing, known to Arab seamen as the "penis of the sea." This particular species — a nearly inert brownish tube — went by the common name of donkey dung. The word *echinoderm* means "spiky skinned," but sea cucumbers as a rule don't have the bristly spines of sea urchins or starfish to protect them. Most have no real body armor and no mobility to get out of the way of trouble. Sea cucumbers move as slowly as glaciers, on little tube feet operated by the hydraulic pressure of their body fluids. They eat sand, taking it in through their mouths, straining it of nutrients in the mysterious hollow chambers of their bodies, and extruding the dross in little turd-shaped piles of sand known as fecal casts. When I picked this sea cucumber up, it contracted a bit, but otherwise gave no indication that it was a living thing. Its skin was thick, but the animal felt squishy and weightless, and it was hard to imagine it being of any nutritional benefit at all to the crabs and other creatures (including humans) that feed on it.

The key component of a sea cucumber is its anus — "this richly endowed organ," as one field guide terms it. To begin with, a sea cucumber breathes through its anus, and some species have gleaming white anal teeth whose function is mercifully obscure. A small fish, the pearlfish, may live inside the sea cucumber's body cavity, using its anus as a door through which to slip out at night. Some sea cucumbers, under attack from predators, will use the opening to fire off a net of toxic threads. The donkey dung has no such projectiles in its arsenal, but if it perceives it is under attack it will create the ultimate diversion by ejecting its own intestines from its handy anus. The attacker — a crab, say — will then eat these sacrificial tissues while the cucumber plods away to regenerate its insides.

I wondered if I could make this donkey dung perform that trick, but my more considerate instincts prevailed and I set it gently back down in the sand. Up above, some of the divers were boarding the boat, and a second group was hovering at the ten-foot decompression stop. Beyond them in the distance were the patient forms of barracuda, dozens and dozens of them grouped together so that their skins flashed like the individual scales of some single gargantuan fish. I sank down beneath an overhang and watched the other divers. When the second group began to board I would swim up to join them, but for now I had plenty of air left to linger. For the moment I considered myself to be in hiding in this coral bower. I lay on my back, watching my bubbles pulse on the ceiling. And through the sunlit cracks came a faint, snowy drift of plankton.

16

The Green Mirror

The sandy lagoon in front of the Island Reef stretched for the greater part of a mile before it hit the solid coral bank of the reef crest. Though I went snorkeling in the lagoon seven or eight times a week, exploring the sand flats and the isolated patch reefs fairly close to shore, I was leery of swimming all the way out to the reef crest by myself. I didn't want to find myself exhausted and alone that far out to sea, in water over my head. But at the end of a day's dive on the other side of the island, I would jump into the hotel pool to rinse the salt off my body and fix my eyes on that distant line of white water. I remembered the intoxicating caves Matt and I had discovered in the coral bank at Algal Ridge, a few miles to the south, and I wanted to see if there were similar catacombs just beyond swimming range of the hotel.

At the pool one afternoon I fell into conversation with Donald, an accountant from the States who said he came to Grand Turk every year to go over the books of a local business. He was in his late thirties, genial and moon-faced.

His glasses were so thick that I caught sight of his eyes only occasionally, drifting into view like fish at the bottom of a deep pond. We exchanged pictures of our children and lapsed into moody talk about the emotional toll of being away from home for weeks at a time.

He was a diver too, he said, and had brought his own gear down, but his work kept him occupied at the times the dive boats went out.

"You and I can go out," I said, nodding my head in the direction of the reef line I wanted to explore.

"We don't have a boat."

"There's one on the beach we can borrow."

I'd had my eye on this boat for weeks, a small aluminum skiff that belonged to the hotel and sat on a nest of turtle grass at the water's edge. No one ever seemed to use it. Arthur had taken its outboard motor into a shed — it needed a new set of coils that might take a week to arrive, or a year — but it could be rowed, and the reef crest was less than a mile away.

Donald and I woke up at five the next morning and hauled our gear down to the beach. Donald's equipment was expensive and new, and I could tell from the unpracticed way he attached his regulator to the tank valve that he had not been out for a while. There was a wind from the northeast, not gusty but persistent, and when the sun rose it illuminated the pitching swells that made the surface of the sea look as corrugated as a plowed field. The swells were not high, only a foot or so at most, but at the reef line the water was lathered and agitated, sending up columns of spume that the rising sun turned as fiery pink as lava.

"What do you think?" I said to Donald.

He shrugged. "We can always come back if it's too rough."

We put our gear into the boat and looked around for the anchor without success. Finally we found a cinderblock, tied it to the anchor line, and shoved the boat out into the water. We each took up an oar and did the best we could with it, but there were no oar locks, just worn loops of nylon line that did nothing to prevent the oars from sawing back and forth along the gunwale as we stroked imperfectly out to sea. Between strokes the wind blew us back to where we had been, and to get any momentum at all we had to row together without missing a beat. I counted out cadence, and we rowed to the rhythm like galley slaves on a trireme. After a half hour we had gone maybe a hundred yards.

Wearily we pushed through the sand flats. One missed stroke would set us back ten yards, and I noticed that the chop was a little higher and that the boat was beginning to ship water. At this rate it would take us hours to reach the reef crest, and once we got there the conditions would be too rough to dive from a boat as small as this one. I spotted the blue shadow of an isolated coral head about twenty yards away and we steered to that. When we got near, Donald tossed the cinderblock anchor over.

"I feel like we're riding kind of low," he said. When we looked down we saw there were three or four inches of water in the bottom of the boat, and more was slopping in all the time, making the little craft as skittish as a canoe. I figured that once we got overboard it would ride higher, and we could grab the anchor line and swim it back to shore. That was no problem, but I felt chagrined at my painfully amateur boatsmanship. Someone who knew what he was doing probably would not have ventured out in these conditions in a creaky rowboat.

Very carefully, not wanting to swamp the boat, we began

to put on our equipment. Sitting in the stern, I slipped on my weight belt, then pulled my b.c. vest, with the tank attached, over my head like a sweater. This was a macho method of gearing up that I had mastered when it had been in vogue a few years earlier, until it resulted in an epidemic of dropped tanks and cracked heads. In a confined space like this slowly sinking rowboat, however, it was just the thing. When I had the tank securely on my back, I put on my mask and fins with the teetering care of a tightrope walker.

In the bow, Donald was almost ready as well. He had his weight belt on and, after some floundering in which the boat took on even more water, had pulled his tank and b.c. vest over one shoulder. Then he stood up to straighten it, and the boat suddenly capsized. As we slid overboard, I gave an amused, exasperated laugh and made a quick mental inventory of all the odds and ends we would have to collect from the water. I looked over at Donald, expecting to see him laughing too, and saw that he was drowning.

His glasses, along with his mask and fins, were still in the boat, so his face was oddly bare and vulnerable, and stark with panic. His b.c. vest, twisted over one shoulder and uninflated, had become a straitjacket, and his tank and weight belt were pulling him down as he grasped at the gunwale of the boat, which was sinking even faster than he was. Looking into his eyes, I saw, in one stark glimpse, how suddenly everything could go wrong. Vividly I recalled Donald's photograph of his children, posed against the mottled blue backdrop at the Sears photo booth, looking into the camera with their trusting, frozen smiles. My mind flashed ahead to dreadful scenarios of myself crawling onto the shore to report his drowning in an uncontrollable, high-pitched voice, or waiting shamefacedly in my room while the tedious search for his body proceeded, or introducing myself to his trembling wife as she arrived at the airport to claim him and his effects.

There was no real reason for him to be in jeopardy, but he was, and I had placed him there. The very equipment that was designed to allow him to exist underwater was now confusing him and robbing him of his strengh. His regulator mouthpiece floated beside him, bubbles erupting from it, and I could see that the boiling water and the sound of escaping air startled him and added to his panic. But the big problem was the boat, which he kept clinging to as it pulled him underwater. I had the shuddering realization that if he went down I would not have the strength or the courage to get him up again.

"Let go of the boat!" I barked.

He shook his head almost wistfully and grabbed it tighter.

"Let go!"

Reluctantly he took his hands off as the boat disappeared beneath the surface. The rapidity with which it went down surprised me, since I had always thought that an aluminum rowboat would float even when full of water.

"Okay," I said to Donald. "Now be calm."

"Yes sir."

"I'm going to take off your weight belt."

"Yes sir."

I slipped underwater and undid the quick-release buckle on his belt, letting it drop to the sand bottom fifteen feet below. Then I came up and pressed the power inflator on his buoyancy compensator, but he started to thrash like a fish as it filled with air.

"Stop it!" I yelled. "Float on your b.c." He looked at me like a man coming out of a deep sleep, said "Yes sir" again, and then lay back on the inflated vest and began to calm down.

I issued a few more stern orders and could hear the panic waning from my own voice. Underwater, my hands were shaking. And something, an unfocused disappointment in myself, was preying on my mind. I should never have blithely assumed that Donald was an experienced diver, nor should I have so seriously underestimated the hazards of a small boat in choppy open water. But the disappointment cut deeper than that: I knew that if Donald had let go of the boat and lunged for me instead, I would have pulled away from him in fright and let him drown.

Supported by our buoyancy compensators, we rested silently on the surface for a moment, and then I went down to the bottom and gathered Donald's mask and fins and the various objects that had been in the boat as it went down. I picked up the weight belt and grabbed one of the oars as it floated by. The other oar was gone, and the boat itself was scudding slowly along the bottom with the current, pulling its cinderblock anchor out to sea.

Donald still seemed to be dreamy with shock, so we let the boat go and snorkeled slowly back to shore. When we got there, Arthur and Melvin were leading another boat through the waves. They had seen us capsize from the beach and had borrowed a wooden skiff whose planking bore only traces of the red and blue paint that had once adorned it.

"Get in," Arthur said, grinning. "We be going to the rescue of the boat."

Donald and I left our gear on the beach and climbed into the skiff. There were three or four conch shells on the bottom and three paddles, one made out of a driftwood limb nailed to a piece of plywood.

Melvin sat in the stern with the makeshift paddle and skulled the boat forward through the building surf while Arthur and I stroked with the other two. Donald sat in the bow, quiet.

"One time at Gibbs' Cay" Arthur said, "Melvin here throw out the anchor without the rope."

"That was you, mon, that done that," Melvin said.

"We had to swim that boat back. That be a *long* swim, *that* day."

Up ahead I saw the prow of the sunken aluminum boat teasingly breaking the surface like a whale's snout. Melvin steered straight for it, but Arthur told him to come off the wind a bit and approach it obliquely.

"Little boat like this, mon, you got to tack, you got to treat it like a sailboat."

"Don't tell me how to steer this boat," Melvin said, "I'm the helmsmon. You just the paddlemon."

The inshore wind kept pressing against us, but the prow of the sunken boat seemed to be heading out to sea.

"We gots to make haste," Arthur said, paddling harder. "It gets in the current we have to chase it all the way to Bermuda. We con stop in Puerto Plata and get some piña coladas."

But we reached the boat before the current took it. It drifted vertically on a shallow bank where the water was not much over our heads, and eventually the four of us managed to heave it onto the surface, bail it out, and tow it in to shore.

Donald stayed around for a few more days to finish his business, then flew back to the States. We had dinner together several times before he left, but we didn't talk much about the capsized boat or his churning panic in the water. I could see he was chewing it over, embarrassed by what had happened, probably resentful that I had witnessed it. I remembered the way he had said "Yes sir" when I had ordered him to stay calm, and how, even in my own fright, I had gloated in the authority he had ceded to me. Although I had almost killed him through my ignorance of the sea's whimsical powers, I managed to convince myself that without me he would be dead, his body skidding along the sand bottom in the clear water of the lagoon.

A week later I was snorkeling in the rain off Governor's Beach. The rain fell lazily in oversized drops that broke on my exposed back like water balloons. At the dock, a hundred yards away, a group of men stood around admiring a new Trans Am that had just been unloaded from a container ship. In the days to come I would see the Trans Am roaring along the pitted roads, restlessly traveling from one tip of the tiny island to the other like a tiger pacing in its cage.

The bottom shelved off deeply here, a bowl of pure sand interrupted only by a few scattered engine blocks and, occasionally, a pioneer clump of coral trying to establish itself. Even with the rain and the overcast, the water was clear, and when I looked down at the bottom twenty-five feet below I felt myself magically suspended above all that space. Like an astronaut looking down at the earth while he soars around it in orbit, I knew I could not fall.

From time to time I would take several deep breaths, bend at the waist, and kick my way to the bottom, cruising for a few moments above the sand before launching myself back to the surface. On the way up I felt the rippling turbulence of the water around my body and the flattening squeezebox of my lungs. Usually I had just enough air left to blow the water out of my snorkel and take a shallow, calming breath. Then I would look down, marveling at how deep I had been.

Once I saw a cushion sea star on the bottom, the sort of bulbous starfish whose dried skeletons are sold in coastal shell boutiques. I wanted to get a close look at the tube feet on its underside, so I dove down, picked it up, and brought it back to the surface. It was a rigid, bulky creature, unlike the squirming brittle stars and basket stars with narrow-gauge arms that I had seen so often on the reef. Its orangish skin was covered with stubby spines and knobs and bright red blotches that formed an elegant pattern. The cushion sea star did not move at all in my hands, though I knew it was alive. I ran my finger over the sucker feet lined up along its five arms, and inspected the mouth on the underside that was the centerpoint of the creature's radial symmetry. Usually cushion stars live in shallower water, where they slurp the bacteria off blades of turtle grass and capture bivalves with their powerful arms. Like many starfish, cushion stars feed inside out, extending their stomach tissues through their mouths and slipping them between the shells of a clam once they have pried it open with their arms. The starfish then secretes enzymes that break down the clam's flesh and allow the roving stomach to consume it.

In its way, a sea star is a vigorous hunter, but the creature I held in my hands was as inert as a rock. I let it drop and followed its progress as it tumbled slowly through the water, an experience that this bottom creeper surely had never had before in its life. It landed upside down, its bland five-pronged underside shining from the depths like a white blossom. A sea star has the ability to right itself in such a condition, but, feeling responsible, I dove down to flip this one over myself.

When I got back, I felt different. Looking down through the water, I no longer had the guarantee that I would not suddenly sink. A moment earlier I had been skimming along on the surface with the confidence of a waterbug, and now I felt the sheer empty space beneath me, pulling on me with slow-acting gravity. What had brought this on? Maybe it was the experience with Donald, which had made me feel for the first time that the ocean was not necessarily my friend, that it could turn on me in a moment. Maybe it was because my time on the island was almost at an end, and I was feeling oddly vulnerable and fatalistic, as if I were as likely to drown as to return to my landlocked home. In any case, the spell was broken, and I saw that it was possible for me to fall through the water as helplessly as that sea star had.

I had never been afraid of water, but now I stroked anxiously with my fins, realizing that there was nothing to support me in this element but my own exertions. I was like a bird beating its wings at the top of the sky. If I stopped moving forward, I would drop. Shore was fifty yards away, and I raced toward it, propelling myself not just with my fins but with my arms, my confidence gaining every moment as I got closer to the beach and the sand bottom curved upward to meet me. Once ashore, I sat down in the rain and grew pensive, wondering if that thoughtless ease I had always enjoyed in the water was gone forever.

What had scared me was all that uninhabited space between me and the bottom, the gathering gloom and density of the water, which for the first time had appeared threatening. It was a fear of falling — or not falling exactly but helplessly sliding downward, without the magic power to swim or scramble back up. Sitting on that beach, I recalled my mother telling me how her own great fear of water had begun. At the age of seven she had walked into the public library of the small Kansas town where she grew up. She had looked through an old-time stereoscope at a three-D depiction of the view from the top of a mountain range and had been frozen with vertiginous fright. After that the thought of water filled her with dread, as if the surface of a lake or an ocean or even a swimming pool was only a veil covering the certainty of an endless free fall.

And yet my mother, when I was fourteen, suggested that I take diving lessons and drove me down to the YMCA to sign up. Why did she do it? "I just wanted you to learn everything about water," she told me when I asked her about it years later, as if water were an enemy that could only be beaten back with knowledge. But maybe it was more than knowledge she wanted to pass on to me; maybe it was a kind of faith. It seems to me now that her decision to turn me over to the element she most feared was not only courageous but insightful, as if she sensed somehow that I needed something larger than myself, and if that something was not to be God and his Catholic church, then perhaps it might be the ocean.

Whether or not the ocean held out the hope of salvation, it was never an enemy to me the way it was to my mother. But I knew it was dangerous, precisely because I had no real fear of it, no serious apprehensions about drowning or sinking.

In the last few days, however, I had had a glimpse of its dark claims. In the folklore of many seafaring countries, people who drown in the sea surrender not only their bodies but their souls as well. Like one of its own stealthy creatures, the ocean lures its victims, then suddenly seizes them and pulls them down into a haunted and unsettled afterlife. Usually, the agent is some sort of intoxicating female apparition — a siren, a selkie, a nymph, a nix, a Rousalki, a White Lady, a gonger, a mermaid. Rarely a mortal human will — like Mary or Elijah ascending bodily into heaven — make the transition to another realm and reside happily as underwater gods in palaces of coral. But usually these humans die, the mermaids feed on their souls or their blood, and their forms — as pale and wraithlike as jellyfish — coast up from the water at night to torment the living, then return at dawn to their gloomy sea caves. To counter the malevolent power of the sea, ancient voyagers painted eyes on the prows of their ships, and later ships bore carved figureheads in the shape of dragons or heroic-faced women whose bare breasts were said to have the power to calm the waves. When sailors were buried at sea, they were often sewn into shrouds; the last stitch was passed through their nostrils to hold them inside, their bodies weighted with chains to keep them from rising again as ghosts.

The rain lasted for only a few minutes, as fleeting as a mountain shower. I watched the cloud that had brought it drift away like a balloon, passing over the container ship at the dock and out toward the sandy point at the southern edge of the island. With the cloud gone, the sunlight poured into the ocean. It acted like a powerful catalyst, scrubbing away the water's shadowy blue tint until all that remained was a colorless clarity. I looked out over the flat surface, thinking of mermaids again. Mermaid legends supposedly began with manatees and dugongs. In waters such as these, sailors would look out over the horizon and see an upright shape rising to the surface, a vaguely human shape with a perceptible head and armlike flippers sometimes clutching a baby to its breast. I never understood how a sailor could possibly mistake a manatee, with its blubbery shape and bristly jowls, for a sea nymph, but apparently something of the sort did happen. In fact, the order to which the manatees and dugongs belong, Sirenia, is named after the sirens in Greek mythology, the water-women whose deadly songs were so bewitching that Odysseus had his men tie him to the mast so he would not be tempted to fling himself overboard. "No life on earth," the sirens sang to him as he writhed at the mast in longing, "can be hid from our dreaming."

The sirens knew that their "green mirror" was irresistible to mortals, and a frequent trait of all such ocean temptresses is complacency and cold-blooded patience. The mermaids of these grim legends are not Disneyesque water sprites with scallop shells positioned coyly over pubescent breasts. They are naked, dripping witches, with a promise in their eyes of some new dimension of sexual rapture. When a mermaid has enticed her doomed human lover to the edge of the sea, she reaches out and seizes him, digging her fingernails into his arms.

The ocean is unknowable, say these tales; it is unreachable. The ravenous mermaids live at the secret heart of nature, in a world they will not and cannot share. Even death will not allow us to enter it.

17

Prospects

In the 1850s a man named J. B. Green, a pioneering salvage diver from New York State, came to the waters of the Turks and Caicos to search for treasure on a sunken British man-of-war. Green had never dived in tropical waters before, though he was a veteran of many adventures beneath the Great Lakes, where he had prowled through various wrecks in his copper helmet and suit of India rubber. Once, when the air hose connecting him to the surface became entangled in forty feet of water, he found it necessary to slice through it with his knife to free himself for a desperate lunge upward. A century before Mike Nelson engaged in underwater combat in "Sea Hunt," Green encountered a diver who was trying to jump his claim on a sunken steamer and beat him off with a crowbar. Another time, Green's partner died a ghastly death when his diving suit collapsed, causing all his blood to flow up into his head, which was still under pressure in the helmet. "On opening the armor," Green wrote, "we found the head very badly swollen, the face and neck so filled with blood as to resemble liver, while the remainder of the body was white as unclouded marble."

When he came to the Turks and Caicos, he found the waters filled with danger and wonder. The British ship Green was searching for had gone down in 1773 carrying eighteen tons of Spanish milled dollars. Though he managed to locate the wreck, Green could not extricate the silver, which was overgrown with coral. But on another wreck he found a great pool of quicksilver shimmering and twitching on the sea bottom, along with a quantity of champagne bottles.

Green and his colleagues wore hoops of iron to guard them against sharks, and when the creatures swam too close the divers disemboweled them with their pikes and watched, amazed, as the sharks fed on their own offal. Acknowledging that he was not a naturalist, Green made a game attempt to catalogue some of the strange and colorful fish he observed on the reef. One of them, he wrote, "resembled the rose in full bloom."

Describing the coral, the former crowbar-wielding Great Lakes diver wrote with uncharacteristic lyricism. He pointed out the towering coral columns rising from the smooth seabed, their labyrinthine interiors, "giving the reality to the imaginary abode of some water nymph."

"In other places," Green went on, "the pendants form arch after arch, and as the diver stands on the bottom of the ocean, and gazes through those lofty winding avenues he feels that they fill him with as sacred an awe, as if he were in some old cathedral."

After his dives in the coral gardens, Green returned to New York, where he was badly bent while trying to retrieve a ship's safe containing $36,000 that lay 160 feet beneath the Atlantic. The accident left Green crippled for life. His career in "Submarine Diving" finished, he wrote his autobiography, published it himself, and sold it to passers-by for a quarter a copy.

I could not get the image out of my mind of this dauntless undersea adventurer, his body ravaged and contorted by decompression sickness, hawking his memoirs on the streets of Buffalo. The wondrous things he saw in the Caribbean would strike most people today as routine. Run through the cable channels on your television, and at any given time you may come across a documentary showcasing the beauty and fragility of coral reefs or the endangered grandeur of the humpback whale. But in Green's time such sights were still largely unknown, even inconceivable. Here was a man who impaled sharks at the bottom of the ocean, who discovered shimmering deposits of mercury, coral temples built on a foundation of lost silver, and underwater grottoes that filled his pragmatic treasure-seeker's mind with sacred thoughts. This man's experiences were so remote from the day-to-day imaginings of his contemporaries that he might as well have been an interplanetary explorer.

Today, even as a mere underwater tourist diving along the Grand Turk wall, I still felt daring and privileged. My sudden fear of the water had subsided, or perhaps had been folded into the deepening humility I felt in the presence of the reef, and now I felt comfortable again as I swam along the wall and flirted with the power of the abyss below. The purity of the light, the intersecting schools of fish, the staggering abundance of the coral, the distant glimpses of roving midwater predators — all of this transported me back into Green's time or, more precisely, into the timeless commotion Green had found waiting for him on the reef when he first descended to it in his bulky armor.

Like many another writer searching for an image to convey

the reef's hushed and otherworldly grandeur, Green had referred to it as a cathedral. Diving on Grand Turk, it was a rare day when I did not feel a twinge of reverence, with the sunlight dancing in shafts all around me and no sound but the hollow exhalations of my own breath through the regulator, a sound so solemn and clinical that in my lightheaded nitrogen musings I thought this must have been what God sounded like when he first breathed life onto the planet.

The reef seemed powerful and eternal. Hidden beneath the waves, it seemed secure from any human depredation. But I knew that it was no more secure than any other aggregate of organisms, no more secure than my own mortal body. One of the divers with us today was the disillusioned owner of a small diving business in the Bahamas. I recognized him underwater by his Ahab-like chin beard and by his easy, expert movements as he slipped through the tunnels and crevices of the wall. The coral bleaching was bad in the Bahamas, he told me, and what was worse was the sheer volume of divers on the reefs. All day long, day after day, year after year, the cattle boats of the big dive operations disgorged hundreds of divers on the same few fragile sites.

"I had thirteen fish that would come up whenever we dropped anchor," he told me on Mitch's boat. "Groupers and snappers, a moray, a barracuda. I started them out on food, I admit, but I didn't keep feeding them. Those fish came up to the divers out of love — to be held and caressed. I went off the island for ten days to see my daughter graduate, and when I came back they were all gone — speared by divers. They've almost destroyed the Bahamas. And when they have they'll march their asses down here and destroy this place."

It is true that the sport of diving began as fish killing. When the skin-diving guru Guy Gilpatric first went down into the Mediterranean wearing goggles, he encountered a three-foot fish known as a *loup*. "Without stopping to think," Gilpatric writes in *The Compleat Goggler*, "I cut loose my right and pasted him square on the jaw." This punching-out of a fish is practically the first recorded act in the history of recreational diving. Underwater, however, bare knuckles were not an effective weapon. Gilpatric soon designed and built a spear, and he astonished everyone by proving that fish could be taken underwater. Soon the shores of the Riviera were swarming with spear-wielding gogglers who would sneak up and stab whatever they came across: *loups, mérous,* octopuses — even electric rays, which sometimes delivered a jolt to the hunter at the other end of the metal spear. Many of the fish had a bovine passivity, but the sport was fair enough. To be successful at all, the divers needed stamina, stealth, and the ability to hold their breath for long periods. But then the harpoon gun came on the market, enabling divers to pot fish from ten to fifteen feet away. Right away, more than a hundred harpoon guns were sold in Nice alone. Gilpatric was disgusted. "Now," he wrote, "even a Shriners' parade can sneak up that close to a fish, so here, at one fell swoop, all the necessity for patience, skill and fish-knowledge was eliminated from goggle fishing."

The invention of the Aqua-Lung added greatly to the divers' already formidable advantage over their prey, and over the decades the big target-sized fish began to disappear from reefs all over the world. Today the situation is better. Spearfishing is as out of fashion as big-game hunting, and at most dive destinations it is not permitted. The last time I saw a diver with a spear gun was in 1973.

There are poachers, of course, and a coterie of fair-minded underwater hunters who still go after fish using only lung power and spears powered by rubber slings. But the real threat to the fish, and to the reefs that sustain them, comes from a multitude of other sources. The coral reefs of the world are dying; we are killing them in all sorts of ways. Though every reef faces its own local hazards, it is also in danger from distant disturbances like ozone depletion or deforestation. Erosion and chemical runoff in one part of the world, for instance, can cause turbidity and suffocating algae blooms in another. Chop down a tree in the Amazonian rain forest, and a coral polyp dies in the Florida Keys.

In the Caribbean, and elsewhere in the world, fish are destroyed, directly or indirectly, by the incalculable thousands. The mangrove forests and spartina marshes that nurture them in their infancy are always in danger of being ruined by oil spills or of being supplanted altogether by the construction of marinas or hotels. Commercial fishermen in pursuit of traditional food fish like groupers or snappers litter the ocean floor with baited fish traps ten feet square. The traps are not selective. They catch groupers and snappers, all right, but also moray eels, parrotfish, angelfish. Very often the traps break away from the buoys they're tied to and remain on the seabed, "ghost traps" that can go on snaring fish for years.

When a coral grazer like a parrotfish dies in one of these traps, the reef's self-defense systems are thrown ever so slightly out of whack. Like many other reef fish, parrotfish eat algae and help temper the natural exuberance of its growth. But when the algae predators start to recede, the algae takes over. In recent years an especially feisty strain of blue-green algae has begun to afflict various species of boulder coral. It starts out possibly where a boring worm, a careless diver, or a boat anchor has nicked a hole in the coral. With time it spreads, choking out the polyps in a distinctive conquering pattern that has given the condition the name of black band disease. Algae thrives on the nutrients found in sewage and in the agricultural fertilizers that our rivers flush down into the ocean. When it becomes overabundant, it can outcompete coral, spreading over the colonies in smothering blankets.

In Southeast Asia, fish are harvested with dynamite and hand grenades, blowing apart the reefs and killing thousands of untargeted denizens with each concussive blast. Fish are often captured for the aquarium trade in the same indiscriminate manner, by dosing coral heads with sodium cyanide. The poison drives the fish out of their holes, killing many of them outright but leaving a few just woozy enough to be captured by hand. At sea, fish and coral are killed incidentally every time a ship flushes the residual oil from its tanks or cleans its engine with caustic chemicals.

I felt the threats to the world's reefs as a personal menace, all the more acutely because I knew myself to be a part of the process that might ultimately destroy them. I was only one of many millions of divers swarming all over them. By and large we are not harmless. We knock over coral formations with careless swats of our fins or rip apart the polyps' external tissues when we touch them with our hands. Although we have mostly laid aside our spear guns, we still pester the fish relentlessly, and we corrupt their natural feeding cycles when we lure them closer with handouts of food. Unless permanent moorings have been installed over the dive sites, the anchors of the dive boats routinely gouge chunks out of the coral. And every new amenity built to serve us — a hotel, a resort, a dive shop, a restaurant, a parking lot — contributes to the outflow of sewage, chemicals, and sediment that could gradually choke the life out of the reef.

Now we were swimming through the Gardens. I had been to this site before, a gradually sloping bank whose low-lying coral surfaces were as dense and compact as a field of gorse. It was shallow here, about thirty feet, and in the pouring sunlight the subtle, velvety hues of the coral had an almost narcotic appeal. According to the *Kumulipo,* the great Hawaiian creation chant, coral was the first living thing, besides man and woman, to be brought into existence. And the coral here did look as ancient as rock. It resembled the cooled and fused remains of some great molten event from beyond the rim of time. Coral, the *Kumulipo* says, sprang from "the exalted Heart of God, from the Blood of Fire, the Almighty Flame of Creation."

It sometimes seemed to me, in my surliest moments, that I had been born into the world just in time to see that flame sputter and die everywhere I looked. In Grand Turk, this was probably a fatalistic overreaction. Except for a few spotty instances of bleaching, a few places where someone had thoughtlessly bombarded the reef with a boat anchor, the corals were still wondrously healthy and undisturbed. That was why, after all, I had come here — to have a glimpse of something that was still pristine, to register it in my memory against the day when the reefs of the world are nothing but silty hillocks of submerged rock. In my heart I believed that day was coming with breathtaking speed, that in the course of one lifetime the oceans would have been used up and contaminated. When my children were my age, would people still be swimming and diving in the sea? Would the waters on the reefs be as clear as they were today, or would they be dark and cloudy with suspended mud?

Writing on my slate now, recording the "kissing" behavior of two blue chromis as they pressed their bony, protuberant mouths together, I stopped in mid-note, suddenly aware that the book I was writing was an invitation for others to join me in overwhelming the reefs of the world with our smiling curiosity. The reef, I knew, would probably be better off without me or my book. The world itself would be better off, say some provocative environmentalists, if humans did not exist at all.

But this was a bitter, scowling, ugly thought. I have a right to be here, I thought as I skimmed over the coral contours of the Gardens. If coral was born from the "exalted Heart of God," then I was too. I was no different from any creature on this reef, except that I had come here with demands and expectations, wanting some sort of revelation that the reef was not interested in providing. But the more time I spent underwater, the more I began to regard that indifference as a sign of acceptance. I watched the schools of wrasse and blue tang motoring across the reef, the eagle rays soaring gloomily overhead, the lizardfish lounging warily in the loose sand. What was I expecting from them? I wanted the impossible: to be noticed, to be invited in, to be taken down into a coral grotto and shown the secret handshake. But there was no secret handshake. *Nobody* was invited. Everybody was just here, no more or less welcome than I. One of the problems humans have always had with the rest of creation is that we insist on seeing ourselves as either superior to it or unworthy of it. What I wanted now was to know what it felt like simply to take my place in it, to accept what it had to offer and to stop pining for gifts of revelation that were beyond my reach.

Down to 900 psi, I rationed my air, taking a breath and then releasing it in a languid, sputtering stream. When I first began to dive, at fourteen, I had sucked in the air in nervous bursts and drained my tank in twenty minutes. Now I breathed through the regulator with as much ease as if I were breathing through gills. But my jaws were sore from constantly gripping the mouthpiece, and my mouth felt the way it did the day after a trip to the dentist. I shifted the regulator in my mouth now and tasted salt water. My throat was dry as usual from the compressed air, my toes were chafed, my mask strap pinched the hair at my temples.

Soon I would be going home, but I tried not to think about that today. I tried not to think at all, but to look at what was in front of me: a porcupinefish, its eyes green and shinily opaque, like crinkled tin foil; a trunkfish, its body no more hydrodynamic than a shoebox, sculling through the surge with its stubby fins; the grains of sand drifting like a vapor across the rills and crests of the exposed sand beds.

I spotted a plastic bag draped across a growth of fan coral and put it in my b.c. pocket, along with the beer can and bottle caps I had collected on a previous dive. Tidying up like this gave me a faint sense of hope and purpose, but my gloomy moods still came and went. What would this reef be like when Grand Turk was a solid phalanx of hotels and resorts, leaking its wastes into the sea? What would it be like after it was ravaged by bleaching or black band disease, or by the bumbling passage of three thousand divers a day in their color-coordinated diving gear? What would the reef be like, I wondered, when it was no longer mine?

18

The Burning Reef

It was my last dive at Harmonium Point. I was feeling restless and worn out, not noticing much, thinking about going home. Mitch and I lined the tanks up along the sea wall as always, then carried them through the lapping surf to the boat. There were only two other divers, a married couple from Canada. The woman was beautiful, with an imperious square face. Her T-shirt said "Hey Mon" in huge block letters.

I kept an eye on the horizon as we headed toward the site, looking for a breaching manta ray or humpback whale, but nothing stirred on the surface of the water, nothing erupted from below. Mitch stood at the stern, nudging the tiller of the outboard with his leg, humming some song against the noise of the motor. The shoreline of the island and the white buildings of Cockburn Town were reflected in his mirrored sunglasses.

When we arrived at the site I helped tie us up to the buoy, then slipped on my gear for the last time and did a backward roll off the boat. I swam down through a school of blue chromis and waited on the bottom for the other divers. A trumpetfish was prowling next to me. I watched it pivot, upend itself, and blow a jet of air into the sand. On a fledgling mound of brain coral, a Nassau grouper perched like some disapproving idol. Its mouth was as wide as a Muppet's, and its black pupils had the shape of bullets.

The Canadian couple were awkward and nervous at first. They had trouble with ear squeeze and leaky masks, and they descended in fits and starts, stopping every few feet to pinch their noses and exchange okay signs with Mitch. When they were settled we coasted along the bottom toward the chasm of Turks Passage. Near the edge of the wall we ducked into a canyon spanned by a thin coral arch and headed downward. I could feel my ears clearing automatically, calibrating the depth — thirty feet, thirty-five feet, forty. Red algae grew along the walls of the canyon, and mossy soft coral marked the place where it opened out into the empty blue of the ocean.

Approaching the end of the canyon, I kicked harder to gather speed and launched myself out into the void. There was good visibility today, a hundred feet, and I could see the wall dropping and shelving off far below me. The big plate corals grew out from it like immense tree fungi, their spherical shapes flattened to better seize the weak, diluted light that spilled onto the reef at that depth. I turned over on my back and spread my arms, looking upward, the sun glinting on my faceplate. Between me and the surface was a swirling, kaleidoscopic mass of fish — blue chromis, blue tang, creole wrasse, silversides. They aroused in me a perplexed longing. Like migratory birds passing high overhead in a silvery sky, they were harbingers of a season's end, of one more measure of time or wisdom that had eluded my grasp.

I saw that the Canadian couple were looking at the fish too, as was Mitch. He lay back as usual, his arms folded across his chest, motionless as a figure on a sarcophagus. After a moment he turned over and led the divers deeper, to show them the black coral growing beneath the promontory at Harmonium Point. They were awkward and cautious in the water, but no longer jittery. They held hands as they swam along the wall, and I noticed that they rarely looked down — and never once looked out toward the blue blankness of the ocean.

Mitch led them along the precipice. He found a lettuce sea slug for them, and a scorpionfish, which he poked into motion with his snorkel. I followed along at a distance, paddling slowly with my fins, then signaled to Mitch that I was taking the cutoff up to Harmonium Point. When I got there and had stationed myself at the forward edge of the promontory, I could see the bubbles of the other divers streaming faintly upward as they proceeded along the wall. The bubbles rose in stringy columns, gaining in size and momentum as the water pressure decreased. At the surface they would have erupted like the waters of a boiling spring. I was suddenly weary of bubbles, these constant outbursts announcing our presence. I suppose I was weary in general. I had gone too far in my underwater quest to turn back, but not far enough to break through. I realized there would be no moment of triumph or transformation, no point at which I would suddenly be filled with wisdom. My craving to melt into the reef, to become an authentic piece of nature, could never be relieved, because in order for that to happen I would have to step outside my human self. The only thing real or true about me, I decided, was my alien watchfulness.

But maybe I was just homesick. It no longer seemed healthy to leave my family for a period of months to chase after a feeling that could not be named or even, finally, experienced. When it came down to it, what could I honestly say about myself except that I was on a search for some rarefied form of sensual gratification? I wanted to be embraced by water.

A spotted moray made its way through the coral shrubbery on the point. I followed it for a while to see what it was up to, but it disappeared into a crevice guarded by a giant Caribbean anemone. Sea anemones are closely related to corals. They are, essentially, solitary polyps, though they usually have many more tentacles than coral polyps and can be huge. This one was a foot across, and though it was attached to the rock with its own cement, it could detach itself and lumber across the reef in search of better prospects. The anemone had more waving arms than I bothered to count. They were greenish with violet tips, as stout as elephant trunks. When I touched one gently with my finger, I could feel its tacky grip.

I wafted over to the ottoman-shaped growth of star coral that sat squarely on the edge of the point. I looked it over carefully for signs of damage or blight, but it was intact, still lush and healthy, the polyps retracted into their green cups, waiting for the failing light to draw them out. A big reef-building coral like this one grows slowly, perhaps a centimeter a year, less than half an inch. In forty years, if I weren't too old to dive down to this spot, it would look much the same: the green coral and the blue water, the sea anemone wriggling its violet-tipped arms. And there I would be, an eighty-year-old man, still looking outward from this undersea mesa, waiting for something to appear.

On the way back, a pod of dolphins surfaced thirty yards from the boat, close enough so that we could hear the percussive sigh of their breathing. Their steely gray backs were the color of the oncoming clouds, and the sight of them in that opaque water made me nostalgic for the Texas coast, where I had grown up watching dolphin fins materialize in silty ship channels and lagunas. Dolphins — porpoises, as we called them then in Texas — had been the emblem to me of all that was unknowable and secret about the ocean, and even now their appearance excited me and filled me with illusions of privilege, as if they had come up to the surface specifically to check on my welfare.

The Canadian couple dug out a camera and scrambled forward to get a better view of the dolphins.

"Oh," the woman said, in a voice that sounded heartbroken, "they're so beautiful."

I hung back a little, not allowing myself to feel the sentimental reactions of the other divers. I had been through my dolphin phase. Perhaps I had read one too many scatterbrained essays about dolphins as the sole hope of the planet. In their natural state, these authors pointed out, dolphins were unfailingly gentle and considerate with one another, were not "hung up" about sex or other bodily functions, followed rigorous ethical standards, and were generally as serene as Buddhas.

I had never let my dolphin reveries take me quite that far, but ever since I was a kid, when I had first read John Lilly's *Man and Dolphin* and swallowed his wild speculations about the possibility of human-dolphin communication, I had bought into the idea that these strange creatures had some special knowledge to impart to me. This, of course, was not the case, but I have never quite gotten over it, and I suppose at the root of my cool detachment on the boat that day was a feeling of unrequited love.

When I was fourteen I tried to catch a dolphin. For a year I had been carrying on a correspondence with one of the directors of Marineland in California, passing myself off as an adult naturalist and writing collegial letters requesting "any information you might be inclined to share with me regarding the maintenance of *Tursiops truncatus* in captivity." Thick reports came back, covering diet, training procedures, gestation periods. I pounced on them with an adolescent's greedy fervor. I never stopped to worry about the way I was abusing my correspondent's good faith, or even about how the dolphin I planned to capture might feel about being taken from the sea. I was a callow, high-minded youth with a job to do.

But as I read through the reports I began to grow restless and disheartened. I see now that I wanted not information but knowledge, and I did not have the patience to carefully sift through the data. After laboring through the reports for a while, my eyes glazing over, I finally set them aside. Thinking through the actual process of the capture, preparing for it, only taxed my spirit. My expert correspondent's data, so generously provided, struck me as irrelevant. Salinity control, fungal diseases, parasites, wholesale suppliers of herring — everything he sent me made the dolphins sound like livestock. But to me they were not even animals, they were *beings*. I sensed that they were waiting for me to approach them, to present myself, and I had every confidence that when the time came, they would recognize me as one of them.

Though I had been researching dolphins for several years with manic thoroughness, I see now that I had no real scholarly interest whatsoever. I simply ignored most of what I read, impatient with any information that did not reinforce the hopeful and bizarre notions that had sprung into my mind. I built a kind of noose, a ring of stiff nautical line padded with foam. My bemused older brother looked at this apparatus skeptically and asked me to explain exactly how it would work. Simple, I told him. I would merely jump into the water with it. It was a well-known fact that dolphins were irresistibly attracted to human beings. They would see me and swim over to inspect me. I would pick out the one I wanted, slip the noose over his snout, and hold on to him while the boat hauled us up the ship channel to his new home, a canal on the back side of Mustang Island, left over from a failed marina development. And who could question that it would work? I was, after all, the authority. In my family's eyes, I think, I was a mildly alarming spectacle, an ungrounded whiz kid in compulsive pursuit of something he could not name. The pursuit could not abide common sense, because what I was after was an ungraspable illusion.

So early one Saturday morning we set out. My father indulgently signed on to pilot our outboard fishing boat through the back bays and into the ship channel that led out into the Gulf. ("If anyone calls," he had said to my mother before leaving the house, "tell them I've gone out to catch a goddam porpoise.") My brother and I wore bathing suits, ready with masks and fins to jump into the water when the dolphins showed themselves. And after awhile they came, heading up the channel toward the boat, a pod of five or six. I grabbed the noose and we slipped overboard, treading water with our fins, dodging the cabbagehead jellyfish that pulsed along just beneath the surface. In the bright morning sunlight the dolphins came closer, arching through the green water, their wet backs glistening, their breath bursting upward from their blowholes in vaporous exhalations that caught the light and created short-lived drizzling rainbows. They were headed straight for me. I held up the noose, ready for them. But then they disappeared, diving deep into the water and not surfacing until they were beyond me, on their way to the blue Gulf.

"I thought you said they'd swim right up to us," my brother said.

I had convinced myself that they would, but as I saw the dolphins swim away I realized that they had taken no notice of us at all. They had not been fascinated. We were just a floating obstacle in their path, something they regarded with routine wariness but no real curiosity. After a casual course correction, they continued on their way, going about their secret business.

We tried again with several other pods of dolphins, and the same thing happened each time. We got back into the boat and headed home, not saying much. I was abashed and angry with myself for my impatience and slipshod planning. But what bothered me more was a deep-seated pang of unworthiness. The true source of my disappointment with the dolphins was not that I had failed to capture them but that I had failed to interest them.

In Cockburn Town that night, they were holding an election rally and talent show. "Miss Turks and Caicos," said an excited young man with a microphone, "will be appearing on stage tonight — *live* — during the progression of our show." I waited around long enough to see Mitch perform "Island Girl" and then left, puttering around town aimlessly on my scooter. On the beach in front of the Kittina Hotel, a group of boys played football, weaving in and out of the waves while streaks of phosphorescence lit up the water.

My plane was leaving at noon the next day. If I dove tonight and flew on a pressurized airplane tomorrow, I would be putting myself at risk for the bends. But I could still snorkel, stroke out through the turtle grass beds behind the Island Reef and look for Lobster Rock one last time. I thought about it as I headed back to my room, but I was tired and anxious to be home, and a bit leery of the dark water. Instead I put all my equipment into the shower and rinsed it off with fresh cistern water. When I had dried every piece with a towel, I put it away in my dive bag and then sat on the porch looking out to sea. From the room next door, the one with a television, I could hear somebody arguing a case on "L.A. Law."

I thought maybe it was all out of my system: that tense expectation of an exalted underwater destiny, the conviction that the reef had the power not only to accept me but to transform me. The fact that this had not occurred, perhaps could not occur, left me neither disappointed nor relieved. I was ready to go home, back to my landlocked life in central Texas, where I knew my scuba gear would sit in the closet for six months or a year before I had the opportunity to use it again. When my children were a few years older, I thought, I would take them down to Mexico, to some sleepy beach resort in Yucatán, where bone-white Mayan temples still stand on the iron shore facing out to sea. I would teach them to snorkel and perhaps gently indulge them if they wanted to learn scuba diving.

For this one night, however, I kept watch, staring out to sea and imagining what was taking place beneath the surface, beyond my sight and understanding. And sometimes I wish that I had not rinsed off my equipment and packed it neatly for the return trip, that I had not been too weary and nervous to enter the water one last time.

Because I still see myself in dreams swimming alone above the rubbly sand flats leading to the wall. It is neither night nor day. The water has a brooding, twilit quality; its blue reaches are soft and deep. Around the isolated coral heads fish are gathered as if in conference, and as I pass by they seem to glance up as if I were just one more flickering item of interest. Ahead I can see Harmonium Point, with its green nubbin of star coral. When I swim up to it, everything is different, charged with expectation. A school of eagle rays swims by in squadron formation, as if announcing some imminent ceremony or procession. Then there is silence, a suspension of movement in the water, an eerie, tactile calm.

Something is approaching, something so huge and powerful that it moves like a great storm, gaining momentum as it seizes all the energy in its path. The advancing pressure waves given off by this creature pound against my body as if I am standing in the surf.

Hovering at the point, I can make out in the distance a feathery strand of white moving slowly up and down. In a moment another one appears, beating in tandem with it, two disembodied angel's wings at the margins of sight. Between them the body of the whale slips leisurely into view. I'm looking at it straight on. It's a featureless wedge — no face, no discernible mouth, no expression, just a knobby point like the prow of a ship cruising out of the fog. The whale's eyes are far back on either side of its head, two glistening bumps that are too small and fixed to animate this blank expanse of flesh. The whale's snout is covered with bumps and tubercles, its pleated underside is as white and smooth as milk. I can hear its groaning songs, mournful pulses of sound that seem to pass through my body like a low-grade electrical shock.

The humpback is close now, thirty or forty feet away, and as it comes to the point where I am standing, it stops and sculls in place with its long white flippers. Something is wrong; it is in some kind of distress or suspense. For a long time it merely hangs there, emitting songs that rumble down into the chasm. The eye that is turned to me is unreadable below its swollen lid of blubber, but the whale knows I am here. It has chosen this place for whatever is about to happen; it has chosen me for an audience.

Its white belly is distended, and when the whale rocks to the side I can see its swollen genital seam encircled by a bristly ring of barnacles. It is about to give birth. I do not record the time it takes for this to happen. I do not worry about the air running out in my tank. There is plenty of air. From time to time the whale contorts herself slightly, lowering her flukes, her graceful body tightening as the calf inside her hitches closer and closer toward the point of its release. Throughout her contractions the whale remains steady, poised at Harmonium Point. A shark skims along in the distance, canny and aware. When I look down at the colony of star coral, I see that it is on fire, a tiny flame erupting from each of its cups.

The whole reef is burning with a votive light. I can see the flames reflected in the white undersides of the soaring rays, and in the brilliant scales of the parrotfish. The sea is darkening, and the luminescent plankton begin to glow in shifting, throbbing colors, until the ocean seems like an arctic sky irradiated by the aurora borealis.

Illuminated by all this dancing light, the whale shudders and sends a blast of song through the sea, the bass notes gently nudging the shark off its course and reverberating through the honeycombed coral rock. The calf's fins appear, unfolding like an umbrella. The fins move slowly up and down, as if the unborn calf is testing their reliability. The mother rests and gathers her strength, and then in one mighty expulsion her baby is free. He is twelve feet long, partly obscured by a cloud of blood and tissue, but I can see the fetal creases still marking his body, and the long umbilical cord still linking him to his mother. Knowingly, he pulls against the cord and breaks free, and then propels himself to the surface, escorted by the mother's white fin.

Through the transparent pane of the surface I can see the calf's first vaporous exhalation. He hangs on the surface for a moment as his mother expels the placenta, which drifts off like a jellyfish toward the circling shark. The newborn's flukes and mottled flippers are soft and limp, but nevertheless he raises his tail and strokes down again, flopping back to his mother, using his bony nose to probe for the twin nozzles of her teats. He clamps one of the nipples in his mouth, and the mother fires off a burst of thick milk. The excess milk drifts like heavy smoke through the water. When he is through nursing, the calf settles in at his mother's billowy underside. She trails her flipper over his body, and sings. The two of them cruise in circles at the end of Harmonium Point. I do not approach them, but I can see the calf's eye looking at me with the same flat curiosity with which he regards the rest of the reef.

Then the humpback whales, the mother and her baby, lift their flukes and descend, following the wall of burning coral, the notes of her song hitting the distant seabed in booming ricochets.

In the dream I take out my slate and write on it — "birth of whale." Now it is daylight, and far above me the sun is high. The coral is no longer on fire, but a plaintive magic lingers. I do not want to leave this boulder of star coral at Harmonium Point; I am as attached to it as the gobies and cleaning shrimp and basket stars that swarm over it, for whom it is their world and their refuge.

An unending school of creole wrasses sweeps across the reef now, blotting out the sun. Each wrasse's blinkered face points dead ahead, an arrow flying to a target I can never see. But the force of their passage uproots me like a powerful wind. Needing to anchor myself, I stroke closer to the coral boulder, and reach out my hand to grasp the living rock.

For Further Reading
Acknowledgments
Index

For Further Reading

My favorite souvenir from my time on Grand Turk is a water-swollen copy of F. Joseph Stokes's *Handguide to the Coral Reef Fishes of the Caribbean* (Lippincott and Crowell). I'm very partial to this little volume, which proved indispensable in my efforts to acquaint myself with the teeming life of the reef and inexplicably brightened my mood whenever I consulted it. It is the best of all the fish identification guides I have seen — compact, comprehensive, easy to use, and as brightly illustrated as a child's storybook.

Stokes's book is admirable but purposely limited. It does not tell you much more about the fish than what they look like, and it does not attempt to catalogue the many other creatures — corals, sponges, worms, mollusks, echinoderms — that share the reef environment. To gain a wider understanding of exactly what a coral reef is, I often reread the excellent essays in two books by Eugene H. Kaplan that are part of the Peterson Field Guide series: *A Field Guide to Coral Reefs* and *A Field Guide to Southeastern and Caribbean Seashores* (Houghton Mifflin). Along with Stokes's fish guide, these two books became part of the gear that I kept in my dive bag.

Of the scores of books that I read on dry land, several stand out as especially valuable or fun. *Undersea Life* (Stewart, Tabori & Chang) combines a clear and authoritative text by Joseph S. Levine with extraordinary underwater photographs by Jeffrey L. Rottman. Rachel Carson's *The Edge of the Sea* (Houghton Mifflin) contains an eloquent chapter on coral reefs. *Watching Fishes: Life and Behavior on Coral Reefs,* by Roberta Wilson and James Q. Wilson (Harper and Row) is an engaging popular introduction to the lifeways of fish. Brian Curtis's *The Life Story of the Fish: His Manner and Morals* (Dover), first published in 1938, is still highly entertaining and coherent. Another book published in that same year, *Animals Without Backbones,* by Ralph Buchsbaum, Mildred Buchsbaum, John Pearse, and Vicki Pearse (University of Chicago Press), still reigns as the classic introduction to the mysteries of invertebrate life. The best field guide and general introduction to corals I've found is the *Ocean Realm Guide to Corals* by Paul Humann (Ocean Realm, Miami).

To learn more about octupuses and squids, the books to read are Frank Lane's *Kingdom of the Octopus* (Sheridan House) and Jacques Cousteau and Philippe Diolé's *Octopus and Squid* (A & W Visual Library). The best-known work on sea turtles — expertly researched and cheerfully eccentric — is Archie Carr's *So Excellent a Fishe,* whose misleading title has been amended to *The Sea Turtle* in the most recent edition (University of Texas Press). I also recommend Jack Rudloe's *The Time of the Turtle* (Dutton).

There is no paucity of reference works or reverential treatises on whales. I found especially useful Richard Ellis's *The Book of Whales* (Knopf) and Robert McNally's *So Remorseless a Havoc* (Little, Brown). For a celebration of whales and dolphins as cognizant beings, take a look at Joan McIntyre's anthology *Mind in the Waters* (Scribners/Sierra Club). Sharks have an equally powerful hold on the human imagination and have inspired nearly as many books. The best browsable volume I know is *Sharks: Silent Hunters of the Deep,* by the editors of *Reader's Digest.*

To learn about the history of diving, I read *Man and the Underwater World,* by Pierre de Latil and Jean Rivore (G.P. Putnam's Sons); *Man Under the Sea,* by James Dugan (Harper & Brothers); and *The History of Underwater Exploration,* by Robert F. Marx (Dover). Jacques Cousteau and Frederic Dumas's incomparable *The Silent World* (Lyons & Burford), is the best-known diving memoir, but I was also inspired by Philippe Diolé's *The Undersea Adventure* (Messner), Guy Gilpatric's *The Compleat Goggler* (Dodd, Mead), and Eugenie Clark's *Lady with a Spear* (Harper & Brothers).

The Encyclopedia of Aquatic Life, edited by Dr. Keith Banister and Dr. Andrew Campbell (Facts on File), is a handy reference for anyone pondering the oceans, as is Cousteau's *The Ocean World* (Abrams), a giant coffee-table book that contains more information about more things than any other single source I encountered. Cousteau, that great compiler, is also the editor (with James Dugan) of another book I grew particularly fond of, *Captain Cousteau's Underwater Treasury* (Harper & Brothers), which is crammed full of adventure stories with titles like "Trapped!" "Surrounded by Piranhas," and "The Phantoms of the Gubbet."

For special information about diving in Grand Turk, consult Captain Bob Gascoine's *Diving, Snorkeling, and Visitor's Guide to the Turks and Caicos Islands.* Newly published, it's available by mail from P.O. Box 101, Grand Turk, Turks and Caicos Islands, BWI.

Acknowledgments

While writing this book, I took advantage of the generosity, expertise, and sheer patience of many people, and it is a pleasure to have the opportunity at last to thank them officially for their kindness. Early on I sought the advice of Ron Coley, an old diving companion with whom I had made many trips to the bottom of the San Marcos River. It was George and Janice Roseberry who first suggested Grand Turk as the locale for this project, and who loaned me various pieces of U.S. Divers equipment to replace my own worn-out gear. Tommy and Shirley Strasberger, in an act of hospitality that was crucial to the writing of the book, gave me the use of a condominium at the Island Reef during my time on the island. I am greatly in their debt and can only hope they feel that the finished book justifies their investment.

On Grand Turk, Arthur Lightbourne provided good company and valuable advice and contacts. Mitch Rolling took me out day after day in his dive boat and helped me to see and understand the reef he knows so intimately. Herbert Sadler distilled for me some of his encyclopedic knowledge of the history of Grand Turk, and Kurt Buchholz and his crew at the Smithsonian research station told me everything I wanted to know about spider crabs and turf algae, and then invited me to Thanksgiving dinner.

At home in Texas, I benefited from the expert guidance of Judith C. Lang, a curator at the Texas Memorial Museum and a contributor to Eugene H. Kaplan's *A Field Guide to Coral Reefs*. Betty Sue Flowers and Ricardo Ainslie helped me understand the psychological implications of the human attraction to water. Donald Keith and his Ships of Discovery colleagues were gracious hosts during my visit to one of their excavation projects in Mexico and provided me with vivid lessons in the reality of underwater archeology. I'm also indebted, for help with logistics and information, to Derek Green, Patricia Sharpe, Suzanne Winckler, David K. Mattila, Richard M. Hammer, Oliver Lightbourne, Gregory Curtis, Cecil Ingham, Bill Wittliff, Victor Emanuel, John Shobe, Robin Burr, Glenn Keller, Craig Quirolo, and William Keegan.

I owe a running debt to my friend Lawrence Wright. Through the years we have criticized each other's manuscripts while gasping through the Hotter 'n Hell bike ride in Wichita Falls, hiking along the Panorama Trail in Yosemite National Park, or wolfing down biscuits at the late, lamented Night Hawk #2 restaurant in Austin. Elizabeth Crook, another writer whose judgment I depend on, provided sound advice and solid friendship. At Houghton Mifflin, Peg Anderson read the manuscript with great care and insight, and Dan Maurer took on many tasks, large and small, for which he has my thanks. I'm also fortunate to have as guiding lights two of the brightest and most committed people in book publishing — John Sterling, my editor at Houghton Mifflin, and Esther Newberg, my agent at ICM.

My wife, Sue Ellen, has gamely accompanied me underwater on several occasions, and she and our three daughters have abided my absences to tropical locales with better humor than I had a right to expect. It was only when I came home to them from Grand Turk that I realized my underwater journeys had been a search for something I had already found.

Index

ATLANTIC OCEAN
FLORIDA
Florida Keys
Grand Bahama
Great Abaco
Eleuthera
Andros
Cat I.
San Salvador
Long I.
Crooked I.
Caicos Is.
Great Inagua
Turks Is.
CUBA
HAITI
DOMINICAN REPUBLIC
Jamaica
CARIBBEAN SEA
Curaçao
Aruba
VENEZUELA
North Caicos
Providenciales
Grand Caicos
East Caicos
West Caicos
THE TURKS AND CAICOS ISLANDS
South Caicos
Grand Turk
Salt Cay
Ambergris Cays
Great Sand Cay